# Electronic Equipment Packaging Technology

Gerald L. Ginsberg

**VNR** VAN NOSTRAND REINHOLD
———————————— New York

*To*
*Dr. Susan Beth Ginsberg*

Copyright © 1992 by Van Nostrand Reinhold

Library of Congress Catalog Card Number 91-30811
ISBN 0-442-23818-5

Manufactured in the United States of America

Published by Van Nostrand Reinhold
115 Fifth Avenue
New York, New York 10003

Chapman and Hall
2-6 Boundary Row
London, SE1 8HN, England

Thomas Nelson Australia
102 Dodds Street
South Melbourne 3205
Victoria, Australia

Nelson Canada
1120 Birchmount Road
Scarborough, Ontario M1K 5G4, Canada

16  15  14  13  12  11  10  9  8  7  6  5  4  3  2  1

**Library of Congress Cataloging-in-Publication Data**

Ginsberg, Gerald.
  Electronic equipment packaging technology / Gerald Ginsberg.
    p.   cm.
  Includes index.
  ISBN 0-442-23818-5
  1. Electronic packaging.  I. Title.
TK7870.15.G56   1991
621.381'046—dc20                                91-30811
                                                    CIP

# Contents

# Preface

The last twenty years have seen major advances in the electronics industry. Perhaps the most significant aspect of these advances has been the significant role that electronic equipment plays in almost all product markets.

Even though electronic equipment is used in a broad base of applications, many future applications have yet to be conceived. This versatility of electronics has been brought about primarily by the significant advances that have been made in integrated circuit technology.

The electronic product user is rarely aware of the integrated circuits within the equipment. However, the user is often very aware of the size, weight, modularity, maintainability, aesthetics, and human interface features of the product. In fact, these are aspects of the products that often are instrumental in determining its success or failure in the marketplace.

Optimizing these and other product features is the primary role of *Electronic Equipment Packaging Technology*. As the electronics industry continues to provide products that operate faster than their predecessors in a smaller space with a reduced cost per function, the role of electronic packaging technology will assume an even greater role in the development of cost-effective products.

This book gives the reader a greater appreciation of the many design, fabrication, and assembly considerations that go into making a successfully packaged electronic equipment device. It also offers approaches to satisfy these requirements so that electronic equipment packaging engineers can make the appropriate tradeoffs when designing a particular product. Hopefully, this book will play its part in continuing the rapid advance of electronics technology.

GERALD L. GINSBERG, P.E.

# 1

# Packaging Implementation

The implementation of electronic equipment packaging concepts into cost-effective and performance-effective end products is a multifaceted undertaking. To varying degrees, it involves the integration of combinations of the basic elements of electronic equipment packaging that will be described in this book.

However, the multiplicity of end product applications, with their specific sets of marketability requirements, makes electronic equipment packaging a very complex task. Thus, the design of an individual piece of electronic equipment or of a system cannot be achieved with the electrical and mechanical design disciplines functioning independently of one another. In fact, the most successful electronic equipment packaging implementation takes into account other nonengineering considerations such as manufacturability.

## 1.1 RELATED GENERAL PACKAGING ISSUES [1]

There are several major issues that have to be considered when beginning the packaging of any electronic equipment. Some of these issues are general and some are very specific as they deal with a particular implementation discipline, performance attribute, or customer need.

### 1.1.1 End Product Requirements

The requirements for the specific end product being packaged vary considerably. Obviously, the packaging engineer must be aware of the specific details within the procurement document or company product line specifications that affect the design of the equipment. However, it is often important for the packaging engineer to be aware of other information, such as budgets, cost goals, delivery schedules, maintainability levels, etc. Although not as obvious, many hardware trade-offs can be significantly affected by these implementation factors.

### 1.1.2  Program Management

At the beginning of a program to develop the end product electronic equipment several specific implementation decisions have to be made for use by the designer. Among the items to be included are:

- System definition
- Functional characteristics and item usage
- Block diagrams of the partitioned elements of the equipment
- Replaceable module definition
- Input/output, (I/O) definition, at each level of partitioning
- Allowances for expansion options, when applicable
- Testability requirements

The company facilities and resources being used on the project should also be defined. When possible, all of these items should be documented in a program management plan with an associated high-level design implementation process flow overview diagram, Figure 1.1.

### 1.1.3  Electronic Design

The need for close cooperation between the electronic and packaging (electromechanical) design efforts cannot be over emphasized. Therefore, it is important that line of communications be established during the early stages of the electronic equipment implementation effort.

A great deal of the success or failure of the program is dependent on having all of the design disciplines aware of the objectives and concerns of the others involved. Thus, as early as the initial "brassboard" and "breadboard" stages of the program, tradeoffs should be made that involve inputs from a broad base of departments within the company, including manufacturing.

Perhaps no single effort is more dependent on this interchange of ideas than is the initial circuit partitioning. It is usual, at this time, that the end product equipment first begins to be formalized into subassemblies whose cost, performance, size, maintainability, etc., most significantly impact the entire implementation program, because at this time, mechanical interfaces are established and the sophistication of the packaging is determined.

It is also important that the electronic packaging designers impart their influence upon other hardware-related aspects of the design that will affect end prod-

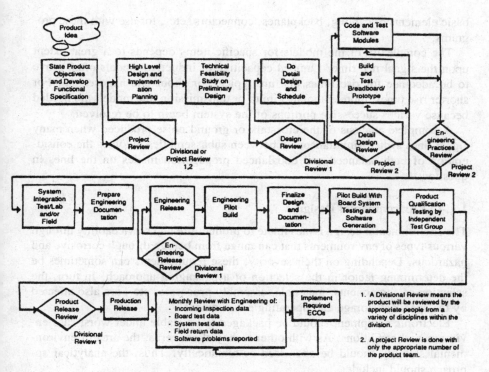

**FIGURE 1.1.**    High-level design process flow overview. [1]

uct performance. For example, the need for appropriate levels of shielding can significantly influence the packaging approach to be taken.

The maintenance of signal integrity and other electrical characteristics throughout the system gain in importance as electronic circuit speeds increase and switching rise and fall times decrease. Thus, in order to realize the performance potential of the newer integrated circuit logic families, the electronic equipment must be designed as a network of "transmission lines." There is also a need for an awareness of related issues, such as controlled impedances, reflections, crosstalk, shielding, and power distribution integrity (see Chapters 5 and 7).

The design of electronic equipment computer programs, such as SPICE (Simulation Program with Integrated Circuit Emphasis), are often used to model and predict electronic equipment performance, including the effects of transmission lines, distributed loads, and terminations. In cases where the electrical properties of subassembly-to-subassembly connections are important to the functioning of the system, it is necessary to develop appropriate models for the basic elements of cabling, backplanes, connectors, etc., for use with these programs.

The complexity of the models for specific items depends to a great extent upon the signal rise time. Lumped capacitive or inductive models often prove to be adequate for rise times that are typical for about one nanosecond. For shorter rise times, more elaborate circuit or transmission line models are needed because various successive portions of the system begin to be resolved.

A complete analysis of the crosstalk or ground noise produced when many signal lines switch simultaneously between subassemblies involves the consideration of both balanced and unbalanced propagation modes on the lines in question.

### 1.1.4  Environmental Design

The environmental design issues relate to product life and serviceability through various types of environments that can range from benign through corrosive and hazardous. Depending on their severity, these requirements can sometimes be the determining factor in the selection of a packaging approach. In turn, the selection of components, materials, assembly processes, etc., are also affected by the specified storage and operating environment.

Electronic equipment should be packaged to be stable under worst-case environmental conditions. As with other technical concerns, the area of environmental stability should be addressed systematically. Thus, the analytical approach should include:

- Defining the basic system functions that are affected by variations in environmental parameters,
- Determining the environmental parameters associated with equipment performance and reliability and,
- Applying cost and producibility factors that can affect the ability of the packaging hardware to contribute toward obtaining a cost-effective end product.

Once the basic product orientation with respect to its environment has been analyzed and related to acceptability criteria, the electronic packaging approach taken can be tailored to optimize the equipment's environmental stability. In the more complex electronic equipment packaging applications, this is often done in conjunction with the use of computer aided engineering technology, such as the system shown in Figure 1.2.

## 1.2  LEVELS OF PACKAGING

Over the years electronic equipment packaging concepts have evolved into concepts that, for convenience purposes, divided the electronic equipment into

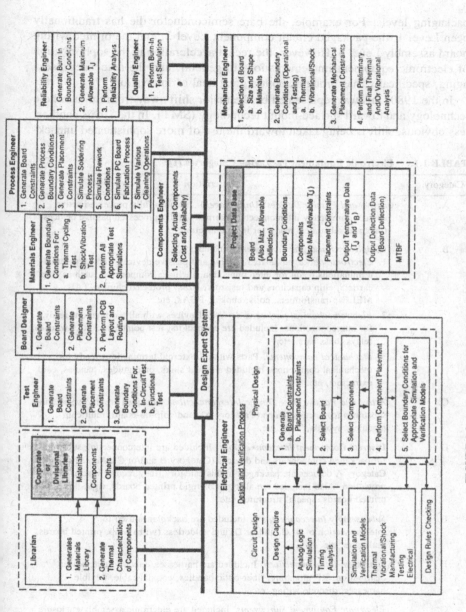

**FIGURE 1.2.** Design expert system. (*Courtesy of Pacific Numerix Corp.*)

5

packaging levels. For example, the bare semiconductor die has traditionally been Level 1, the packaged circuit component Level 2, and the printed wiring board assembly Level 3. However, the recent acceleration in the sophistication of electronic equipment implementation, in general, and the associated packaging, specifically, has brought about a great deal of innovation.

In the 1980s this was represented by the major shift away from through-hole technology and toward surface-mount technology (SMT). In the 1990s another, less obvious, shift is being taken toward the use of more sophisticated unpack-

**TABLE 1.1.    Electronic Equipment Packaging Categories [1]**

| Category | Description |
|---|---|
| A. | *Active and Passive Devices (Unpackaged).* Includes fully processed active devices completely interconnected internally; fully processed passive devices (inks and substrates) ready for interconnection. |
| B. | *Completely Packaged Components.* Includes:<br>1. *Electronic packaged devices.* Active and passive devices with all external terminations ready for interconnection. Includes components, such as chip carriers, chip capacitors and resistors, DIPs, protected chip-on-TAB, MELFs, transformers, coils, chokes, PGAs, etc.<br>2. *Electromechanical packaged devices.* Devices with all terminations ready for interconnections. Included are connectors, test points, sockets, switches, relays, buss bars, etc.<br>3. *Mechanical components.* Parts with all external terminations ready for mechanical connection. Included are heat sinks, card guides, handles, card extractors, etc. |
| C. | *Active and Passive Base Device Interconnections.* Included are interconnecting substrates for uncased chips (Category A), and chip-on-board, multichip modules, hybrids, TAB, etc. |
| D. | *Discrete Component Interconnections.* Included are interconnecting substrates for packaged components and devices (Category B and/or C). *Note:* If Category A devices are packaged in this category, they require special protection. Also included are all types of rigid printed boards, rigid-flex printed boards, special structures, etc. |
| E. | *Subassembly Interconnections.* Included are backplanes used for interconnection of B, C, and/or D, and solderless (wire) wrap, printed boards, multiwire, etc. |
| F. | *Subsystem Intraconnections.* Included are harnesses, input/output, power distribution, RF plumbing, fiber-optic bundles, coaxial cable, flexible harnesses, discrete wiring, etc. |
| G. | *Electronic Equipment Subsystems.* Included are electronic assembly enclosure (black box), and the input/output hardware, card racks, molded handsets, etc. |
| H. | *Integrated Electronic Assemblies and Systems.* Includes all end product hardware for aircraft, missiles, computer systems, antennas, telecommunications, etc. |

aged integrated circuit assemblies as represented by the emerging fine pitch technologies (FPTs) that are based on the use of chip-on-board (COB), multichip module (MCM), and tape automated bonding (TAB) processing.

With this in mind, the Institute for Interconnecting and Packaging Electronic Circuits (IPC) deemed it appropriate to formally establish definitions for the levels of electronic equipment packaging that relate to the various implementation options that are available. Toward this end, the IPC established eight categories (Table 1.1), and eight functions (Table 1.2) of electronic equipment packaging. These categories and functions form an 8 x 8 matrix, i.e., there is a function that is applicable for each category, and vice versa.

**TABLE 1.2.    Electronic Equipment Packaging Functions [1]**

| Function | Description |
|---|---|
| 1. | *Design*. Deals with the design rules and the requirements for products, the electrical and physical properties and parameters. Mechanical attributes, testability, thermal requirements, and software concepts are evaluated for the environmental considerations of the packaging category. |
| 2. | *Data and Information Generation Transfer*. Deals with the language and format standards, as well as the media and protocols needed for transferring information, either electronically or via magnetic tape. Interface and intraface descriptions are identified. |
| 3. | *Base Materials*. Deals with both dielectric and conductive materials and substrates. Included are foil, plating, inks, coatings. This function also deals with the metal support structures necessary to build multilayer printed boards, as well as process consumables. |
| 4. | *Fabrication Processes*. Includes all types of processes, such as dry, liquid, mechanical, and special. Describes the automation and process controls necessary to achieve high first pass yields. |
| 5. | *Assembly and Joining Processes*. Deals with both automatic component placement and insertion, as well as the handling of these parts. Attachment and joining techniques are described, as well as the cleaning operation prior to coating and encapsulation. |
| 6. | *Product Performance*. Deals with the design/performance/assembly and testing criteria for all types of products, such as rigid printed boards, flexible and flex-rigid printed boards, metal core boards, backplanes, hybrid circuits, multichip modules, discrete wiring, and electronic assemblies. |
| 7. | *Product Assurance*. Deals with the qualification, quality assurance principles, and test methods. Also responsible for process control management and developing techniques for evaluating statistical process control verification of product performance. |
| 8. | *Management*. Deals with the initiation and status reporting on various processes involved in the electronic packaging arena. Establishes control parameters and profitability aspects of managing the business aspects of making equipment. Also included are statistics and marketing strategies. |

## 1.3  DESIGN IMPLEMENTATION TEAM [1]

It is important to establish a close-knit design team in order to successfully develop an electronic product. Depending on the product being implemented, the following disciplines should be involved in the design process.

### 1.3.1  Systems Engineering

Systems engineering is the group responsible for determining how the contract, customer, or marketing requirements are transformed into end product circuits and hardware. Unfortunately, the importance of this discipline is often overlooked or ignored. However, without a firm set of clearly defined end product implementation requirements, all of the subsequent efforts on the program will be less than optimal.

### 1.3.2  Circuit Design

The role of the circuit design effort is to begin the conversion of system requirements into distinct assemblies and subassemblies. This is a particularly critical effort in the implementation cycle. Thus, it is important that the output of this effort, in the form of logic diagrams, schematics, parts lists, etc., clearly define all aspects of the circuit design that are necessary for the end product to perform reliably. If the results of this effort are subject to continual change, it can be expected that the remainder of the implementation effort will be exceedingly difficult.

### 1.3.3  Software Engineering

The importance of software engineering is increasing in significance as electronic products become more closely linked to data processing (computer) operations. The output of this effort is another set of requirements that will affect the other disciplines. Therefore, sufficient time and effort should be made available for this function during the early stages of the product implementation activity.

### 1.3.4  Packaging Engineering

Being the subject of this book, packaging engineering is the primary discipline through which all of the circuit design information flows and through which a feasible product can be designed and documented. In addition to the mechanical aspects of this discipline, packaging engineering either includes or provides the major interface with the disciplines responsible for printed board design, component selection, standardization (producibility), reliability, etc.

### 1.3.5 Computer-Aided Technology

In some companies the design, fabrication, assembly and testing efforts are supported by computer-aided technology specialists. With respect to electronic equipment packaging, when implemented properly, the use of computer-aided design (CAD) can significantly assist in performing structural, thermal, and environmental stability analyses.

### 1.3.6 Printed Board Design

Printed (wiring/circuit) board designers, as well as those responsible for discrete interconnection wiring cabling, are important members of the design team. Since this is one of the significant steps in the product implementation process, this "nonrecurring engineering" effort will greatly impact "recurring" fabrication, assembly and testing costs. Thus, maximum benefits will be accrued if sufficient time and effort is allocated to this discipline.

### 1.3.7 Component Engineering

Component engineering can play a particularly important role when advanced electronic equipment technologies are being implemented. This is because the availability of new electronic components is constantly changing and their proper selection not only affects circuit performance, but also impacts producibility, reliability, maintainability and other life cycle costs.

### 1.3.8 Standards Engineering

Standardization is a discipline that can play an important role in the implementation of an electronic product. When used properly the contributions from standards engineering personnel can yield definite cost savings benefits to the project.

### 1.3.9 Reliability/Maintainability Engineering

The role of reliability/maintainability engineering is gaining in importance for the development and marketing of many electronic products. This is because of the increasing awareness of these disciplines by the product consumers and the increased usage of electronics in performance critical products such as life-support applications.

## 1.4  PRODUCT APPLICATIONS

The first thing that an electronic equipment packaging engineer should do at the beginning of a new product development program, regardless of its size or complexity, is to visualize equipment as a "system." Thus, the packaging engineer should acquire as much knowledge and understanding as possible of the product's end use and life cycle environment.

It will then become apparent that the optimum approach to the packaging of the equipment vary depending on the specific requirements of the end product application. This is because each of the basic categories of end product applications have their own unique considerations and, in fact, within each group there will probably be many variations of these requirements.

### 1.4.1  Consumer Electronics

Perhaps more than with any other type of equipment, consumer electronic products have to be packaged with the end product user in mind. Some of these units may be suitable for use in "business" environments and, conversely, some business products are used by individual consumers. Thus, there can be an overlap of requirements.

The functional requirements for consumer electronics should relate to how the product must perform, i.e.:

- What features of the product most affect its appeal to the consumer (marketability)?
- What role does reliability play in the performance goals for the product?
- How are the end product and life-cycle costs requirements?
- To what extent must cosmetic factors be taken into account?

In addition, the environmental conditions under which the product must function (or be stored) have to be clearly delineated.

The use of electronic products by the consumer in the home is becoming commonplace. In addition to appliances, electronics are used extensively in entertainment, communications, data processing and other household applications.

In some instances the home environment can be more severe than the office environment. For example, automatic electronic garage door openers need to operate over a wide range of temperature and humidity extremes.

Of course, safety is also of prime importance. Quite often the equipment must be packaged in a manner that prevents inadvertent exposure to high (lethal) voltages. Thus, if user serviceability is required for these products, appropriate seals and interlocks will have to be provided.

Home product purchasers are often very price conscious. Also, although reliability and quality are important, style and appearance are often overriding marketing considerations.

### 1.4.2 Business Products

The use of electronic products has increased significantly in the business (office) environment. Since the business environment is relatively benign, the demands on the equipment sometimes barely exceed those associated with the comfort zone for office personnel. However, when cabinet rack-mounted equipment is used it often has to provide for its own thermal management (forced-air cooling fans and blowers) and shielding gasketing.

Because of the Occupational Safety Hazard Act (OSHA) requirements, the control of ambient noise is an important packaging consideration. In a similar manner, Underwriters Laboratory (UL) standards for product safety need to be taken into account. Federal Communications Commission (FCC) limits for electromagnetic interference (EMI) and radio-frequency interference (RFI) create a unique set of concerns for the electronic packaging of the equipment.

The physical appearance of the equipment can affect the marketability of business and consumer products. Thus, to varying degrees, it is important that these products outwardly convey an image of quality.

### 1.4.3 Mobile Products

The use of electronic products in automobiles, aviation and marine applications brings about a wide range of environmental extremes that are unique to this type of equipment. In addition to the traditional concern with shock, vibration, and temperature, the packaging of mobile electronic equipment also involves the simultaneous involvement with rain, salt spray, fog, humidity, sand and dust.

Automotive products located in the engine compartment may be exposed to the worst combination of environmental stresses than is any other electronic equipment, including those for military applications. Products located in the passenger compartment are less severely affected, but thermal considerations may significantly influence the packaging of the electronics.

Aviation and marine products also have unique environmental stability requirements. For example, the thermal management of aviation equipment can be significantly affected by the reduction in convective cooling at high altitudes where the air density and pressure are lower. The exposure to corrosive salts in shipboard products can also be significant.

Product size and weight limitations can be major mobile equipment packaging restraints, especially in aviation applications. Reliability, maintainability and man-machine interface requirements are often major elements of a mobile equipment packaging tradeoff analysis.

### 1.4.4   Industrial Electronics

The wide variety of industrial factory environments means that the electronic products storage and operating environment must be clearly defined. Such environments can vary from benign clean rooms to areas where etching, plating and cleaning chemicals are used. When appropriate, the use of the equipment in potentially explosive atmospheres can override most other packaging considerations.

Depending on the application, OSHA, UL and FCC requirements can be applicable to this type of product. Thus, it is not uncommon for such products to be packaged in cabinet racks or National Electrical Manufacturers Association (NEMA) enclosures that are weather resistant and oil tight.

### 1.4.5   Medical Electronics

For obvious reasons medical electronics, especially life-support equipment, must be of the highest reliability; size and weight are also often major considerations. Thus, medical electronics packaging can range from small hermetically sealed implantable modules to desktop enclosures.

### 1.4.6   Military and Space Electronics

The packaging of electronic equipment for military and space applications usually are required to comply with a wide range of specifications and standards. Thus, unlike some of the other product types, all of the requirements for size, weight, environmental stability, reliability, maintainability, etc., are clearly defined.

Many of the requirements for this type of electronic equipment can be found in MIL-STD-454 which is an omnibus document that refers to various standards and specifications that ultimately affect the packaging of the end product. In many instances this includes the procurement of components and materials from Qualified Product Lists (QPLs), and the fabrication, assembly and testing of subassemblies and assemblies in accordance with the requirements of detailed specifications.

## 1.5   PACKAGING PRIORITIES

It is import that the electronic equipment packaging engineer be cognizant, at the beginning of the equipment implementation effort, of the diverse factors that can influence the final product. In addition to the general considerations just discussed for generic groups of electronic products, a weighted list of priorities should be developed for the requirements of the specific equipment being designed, such as those shown in Table 1.3.

### 1.5.1  Hardware/Software Relationships

The relationships between electronic hardware and software are generally determined by the engineering team prior to the beginning of the formal electronic equipment packaging phase of the program. The determination of what end product functions will be performed in the form of hardware or software is a decision that must take into account more than packaging, although packaging is an important factor.

Support and maintenance criteria should also be considered in the allocation, location and partitioning of software and hardware functions. This is particularly significant where the software and hardware relationship becomes what is referred to as firmware.

### 1.5.2  Firmware

Firmware is typically software that resides in a hardware medium, such as packaged programmable/erasable memory devices that are mounted on printed wiring board assemblies. Thus the use of firmware is a viable part of a product

**TABLE 1.3.   Typical Electronic Equipment Packaging Priorities [1]**

| Factor (In Alphabetical Order) | Application | | |
|---|---|---|---|
| | Consumer | Industrial | Life Support/ Military |
| Appearance | 6 | 10 | 10 |
| Brand image | 10 | 5 | 6 |
| Human factors | 12 | 7 | 5 |
| Life cycle costs | 7 | 1 | 3 |
| Maintainability | 9 | 2 | 2 |
| Market share/competition | 2 | 8 | 9 |
| Price | 1 | 6 | 7 |
| Promotion/public relations | 11 | 12 | 12 |
| Quality/reliability | 4 | 4 | 1 |
| Quantity | 3 | 9 | 8 |
| Styling | 5 | 11 | 11 |
| Warranty/customer support | 8 | 3 | 4 |

Number 1 is the most important and number 12 is the least important factor.

implementation strategy. However, careful attention should be given to partition and allocation of software functions in order to achieve practical and economic support and maintenance objectives. Future enhancements and performance upgrading should also be taken into account in the software/hardware/firmware allocation.

The software language should be a matter of concern for a broad spectrum of influences, just as the hardware/software interfaces are. This is because the selection of a software language will help to determine many features of a product. The language selected can influence the type, size, and availability of implementation hardware, technical and functional capabilities, support and maintenance costs, operational flexibility, performance, user friendliness, and, thus, product marketability.

### 1.5.3  Hardware Description Languages [2, 3]

Several standard computer-aided design (CAD) languages already have been developed which can be used to describe various electronic equipment packaging hardware and other design functions. Some of these languages have not necessarily been developed for printed circuit board functions, but because of their applicability to other facets of the design process, they can also be used in printed circuit board applications. These include such standards as:

- IGES/PDES
- EDIF
- VHDL
- IPC-D-35X

Their interrelationship is shown in Figures 1.3 and 1.4. In addition, users may have their own native data format standards. These are sometimes related to the vendor equipment purchased by the user.

Some companies have standardized the various departments within their organization in order to allow for electronic communication of product data. These are usually proprietary formats and not industry standards.

Such digital descriptions are desired in order to facilitate the automation process for producing parts. On-line electronic data transfer is possible and intended to:

- Eliminate the need for human intervention,
- Facilitate the storage of documentation in a format other than paper (archiving)
- Create a standard way of describing design data that can be transported to machines other than the one on which it was created.

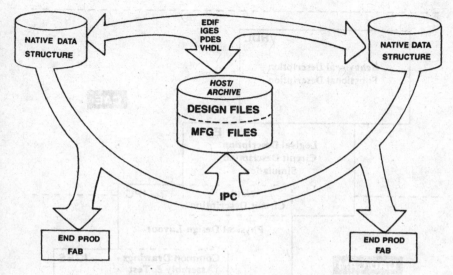

**FIGURE 1.3.**    Electrical data structure interfaces, [2]

### 1.5.3.1   Initial Graphics Exchange Specification (IGES)

The Initial Graphics Exchange Specification (IGES) is a communication file structure for data produced on and used by Computer-Aided-Design (CAD) and Computer-Aided-Manufacturing (CAM) systems. It has been designed to serve as a receptacle for the data generated by commercially available interactive graphics design drafting systems. This structure provides a common basis for the automation interface.

IGES information models have been developed for specific application areas, mechanical products, electrical products, architecture, engineering and construction, finite element modeling, drafting, constituent technical areas, manufacturing technology, solid modeling, curve and surface modeling, and presentation data.

The methodology for data storage of IGES is an entity attribute database. Each record in the file associated with an entity contains parametric data related to that type of entity. Any entities that have a dependency relationship with other entities have within their records a pointer that defines that relationship.

Extensions of IGES will include representation of the following design related information:

- Integrated circuit design and connectivity
- Testing
- Simulation

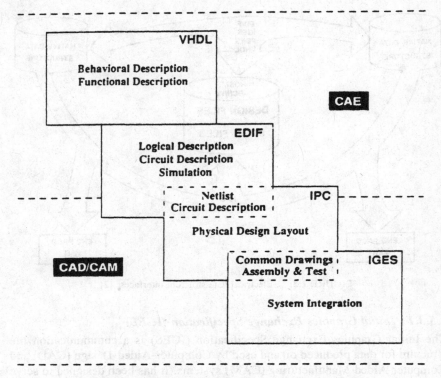

**FIGURE 1.4.**    VHDL, EDIF, IPC and IGES relationships. [2]

- Inspection
- Hierarchal electronics system design.

The IGES organization is also involved with developing the Product Data Exchange Specification (PDES). This research and development effort addresses a different technology base than does the IGES format. IGES is intended for information exchange among data bases that must be interpreted by human beings.

### 1.5.3.2  Electronic Design Interchange Format (EDIF) [4]

The Electronic Design Interchange Format (EDIF) facilitates the movement of electronic design data among sophisticated databases in a commercial environment. This provides links between disparate systems as well as corporate control over information content. EDIF is intended for the representation of integrated circuit (IC) and electronic design data, not mechanical design data.

EDIF is capable of representing the electrical characteristics necessary for implementation of electronic products. It can express library/cell organization; provide extensive data/version control; describe the cell interface, cell details (contents), and technologies; and represent timing, geometry, and physical objects. It also addresses the integrated circuit description necessary for fabrication. This includes modeling and behavioral aspects and some printed board descriptive material.

The EDIF printed circuit layout view is intended to include all of the physical design information for a circuit board, including traces, lands, targets, holes, planes, and vias. In addition, it incorporates all of the part and connectivity data that can be specified in the netlist view.

Standard EDIF allows unlimited flexibility in the use of hierarchy and other grouping mechanisms for designs and subdesigns. Some of the hierarchical levels may correspond directly to physical objects such as components, pins or lands. Other levels may exist only for convenience or to compactly represent data that are shared among different objects. However, EDIF makes no distinction between convenience groupings and real design objects.

### 1.5.3.3  VHSIC Hardware Description Language (VHDL) [3]

VHDL, the Very High Speed Integrated Circuit (VHSIC) hardware description language was approved as an IEEE standard (1076) in late 1987 and mandated by the Department of Defense under MIL-STD-454, Revision L, for application-specific integrated circuits (ASICs) in early 1988. As such, it makes a significant impact on both integrated circuit and printed circuit board design.

The specific purposes of VHDL are:

- Provide a standard medium of communication for hardware design data
- Represent information from diverse hardware application areas
- Support the design and documentation of hardware
- Support the entire hardware life cycle.

A fundamental tenet of VHDL is that the design specification should remain independent of its implementation. This pivotal rule focuses on the reusability of parts and ease of maintenance. The dedicated emphasis on commonality and reusability is structured to reduce the maintenance costs and effort demanded of manufacturers when parts are unavailable or when design revisions are needed.

VHDL is not the easiest design language to learn and apply for any specific function. Thus, there are instances in which another language may be more appropriate. However, for most designs, VHDL permits users to replace proprietary languages and apply topdown methodologies. Thus, it results in a system design that emphasizes behavioral aspects of a circuit or printed board.

The major portions of a VHDL description are entities, architectures and configurations. An entity is a design block or component and, at its lowest level, could be a single Boolean gate. An entity declaration includes the number, type and direction for each type of pin along with a series of parameters, called Generics, that are passed to the component model, called the architecture. Entities can be thought of as schematic symbols, architecture as the simulation model for a symbol, and configurations as the link that associates the entity with one or perhaps several different architectures.

The packaging facilities inherent in the VHDL language, along with the features that permit abstraction of data, allow the user to unambiguously specify the way a design should function. Thus, the VHDL simulator provides very high quality feedback, optimizes design efficacy, and advances the usability of computer-aided engineering (CAE) tools for hardware designers.

### 1.5.3.4  IPC-D-35X Series [5]

The Institute for Interconnecting and Packaging Electronic Circuits (IPC) has generated a set of documents, the IPC-D-35X series, that was developed for the purpose of digitally describing the logical and physical elements necessary as input to a design system. A part of this description is the network or interconnection of physical and electrical parameters among the various electronic parts.

The IPC-D-35X series of standards were developed to specify printed board data in a machine-independent digital format for communication from design to production. This series has evolved to encompass fabrication, documentation, assembly and testing. The format supports data communication among computer-aided engineering (CAE), design (CAD) and manufacturing (CAM) systems.

The IPC series of standards started with IPC-D-350 which is used to describe the fabrication and artwork of a printed circuit board. It has grown into a series of standards that describe the printed circuit board in many different ways, including the schematic diagram, assembly drawing, electrical description, test data, etc. These standards also have the ability to describe the systems in which the boards reside.

The use of the IPC-D-35X series is flexible as it allows the user many options. The structure of the standards allows for the addition of new information and new concepts. This is done by developing new data information modules that describe a particular parameter or facet of the design. As future enhancements are required, the concepts used in the IPC-D-35X series can have record formats added to assist in describing parameters needed by the design community.

The IPC-D-35X series has been developed with the capability for identifying a change in language in midstream. Thus, in a particular data file for a specific printed circuit board, the information might start out in a IPC-D-350 format, switch to IGES, switch back to IPC-D-350, switch to EDIF, and finally conclude in IPC-D-350.

This manipulation of data languages is intended to enhance the use of IPC-D-350 as compared to the other languages. An example of how such an engineering and design system could be configured is shown in Figure 1.5. Also illustrated are the types of application tools for which the data standards are used.

Presently, other languages do not easily identify the return to IPC-D-350, if that is required. This must be checked by the user in order to make a determination on how to move from other languages back to the IPC-D-350 format.

The series consists of:

A. ANSI/IPC-350, Printed Board Description in Numeric Form. The most-commonly used specification for transferring printed circuit board data (artwork, phototooling and physical layout) from one computer environment to another. The data is in a human readable, 80-column card,

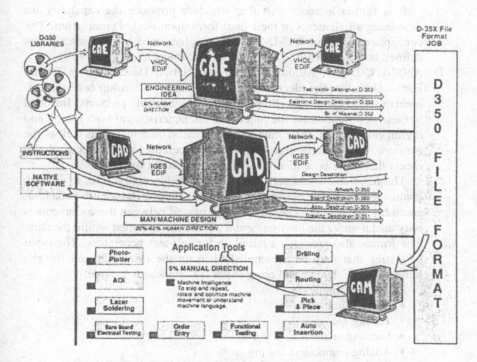

**FIGURE 1.5.**    Typical IPC-D-350 design system interfaces. [2]

ASCII format. Also, IPC-D-350 data can be prepared manually using a digitizer or from more-sophisticated CAD equipment.

B. ANSI/IPC-D-351, Printed Board Drawings in Digital Form. The supplement IPC-D-350, specifically in the area of describing printed circuit board related drawings in digital form. Both of these specifications have been adopted for use by the Department of Defense. The basic record formats in IPC-D-350 also apply to IPC-D-351.

IPC-D-351 pertains to four basic types of drawings, i.e., printed circuit board schematic diagrams, printed circuit board master drawings, printed circuit board assembly drawings, and miscellaneous part drawings.

C. ANSI/IPC-D-352, Electronic Design Data Description for Printed Boards in Digital Form. Describes the relationship between the elements used in the electromechanical design and packaging of electronic products using printed circuit boards as the major form of interconnection. Included in these descriptions are the logical and physical elements necessary as input to a design system, as well as the network or interconnection description between the various electronic parts.

It is further intended that this structure provides the capability for describing all elements in their final form upon design completion. The concepts detailed in IPC-D-352 are supplemented by the descriptions defined in the other IPC-D-35X series documents.

D. ANSI/IPC-D-353, Automatic Test Information Description in Digital Form. Describes the electrical and functional relationship between elements used in the design and packaging of electronic products. Included in these descriptions are the functional characteristics of both analog and digital components, their state and electrical requirements, as well as the network interrelationship descriptions between the various electrical elements that make up a functional electronic circuit or assembly.

The functional characteristics are described in digital form in order to enable data exchange between systems that support design, manufacture, assembly and testing. The information passed between these elements is only useful under the environmental conditions specified within the data. The format also provides a template for self-test generators. The types of testing that may be accomplished using the principles and formats defined in IPC-D-353 and companion IPC standards include:
- Analog simulation
- Digital simulation
- Timing analysis
- Loading analysis
- Analog component testing
- Digital component testing

- In-circuit testing
- Bare-board optical testing
- Burn-in

In addition, IPC-D-350 may be used for bare-board electrical testing and bare-board optical testing

E. ANSI/IPC-D-354, Library Format Description for Printed Boards in Numeric Form. Supplements the other standards in the IPC-D-35X series. It describes the use of libraries within the processing and generation of information files. The data contained in IPC-D-354 covers both the definition and use of internal, exist within the information file, and external libraries.

The libraries can be used to make generated data more compact, and thus facilitate data exchange and archival. The subroutines within a library can be used one or more times within any data information module. As a supplement, IPC-D-354 does not cover the specifics of the other standards in the IPC-D-35X series.

F. IPC-DG-358, Guide for Digital Descriptions of Printed Board and Phototool Usage per IPC-D-350. Enhances and elaborates on the concepts detailed in IPC-D-350. It provides detailed step-by-step examples in all facets of using IPC-D-350 in order to define printed circuit board geometry and other vital characteristics. It also explains the flexibility of implementation and levels of IPC-D-350 to give the user a better understanding of the options provided.

IPC-D-358 identifies those portions of IPC-D-350 that are machine processable. Another purpose is to identify those portions that are intended to provide insight for the user who may receive a IPC-D-350 file and needs to produce a printed circuit board without any further instructions in paper form.

G. ANSI/IPC-NC-349, Computer Numerical Control Formatting for Drillers and Routers. Defines a machine readable input format for computer-numerical-control (CNC) drilling and routing machine tools related to the printed circuit board industry. The format may be used directly to transfer drilling and routing information among printed circuit board designers, manufacturers, and users as the output CNC standard from converters that expand higher-level design input, such as IPC-D-350 data.

## 1.5.4 Computer-aided Acquisition and Logistical Support (CALS)

The Computer-aided Acquisition and Logistical Support (CALS) program is a Department of Defense initiative to enable and accelerate the use and integration of digital technical information for weapon system acquisition, design,

**TABLE 1.4. Numerical Design Process Functional Matrix [1]**

| Design Process | System | | | | Box | | | | Printed Board | | | | Component | | | |
|---|---|---|---|---|---|---|---|---|---|---|---|---|---|---|---|---|
| | IGES | EDIF | VHDL | IPC | IGES | EDIF | VHDL | IPC | IGES | EDIF | VHDL | IPC | IGES | EDIF | VHDL | IPC |
| **A. Behavioral description** | | | | | | | | | | | | | | | | |
| 1. General | 2000 | 5000 | 4311 | 3000 | 2000 | 5000 | 4311 | 3000 | 2000 | 5314 | 4111 | 5102 | 5311 | 4111 | 4111 | 5102 |
| 2. Quality level | 2000 | 5000 | 4311 | 3000 | 2000 | 5000 | 4311 | 3000 | 2000 | 5314 | 4111 | 5102 | 5311 | 4111 | 4111 | 5102 |
| 3. Signal | 0000 | 2100 | 5533 | 3000 | 0000 | 5322 | 5533 | 3000 | 0000 | 5322 | 5533 | 3000 | 5422 | 5533 | 5533 | 3000 |
| 4. Ports | 0000 | 2100 | 5533 | 3000 | 0000 | 5322 | 5533 | 3000 | 0000 | 5322 | 5533 | 3000 | 5422 | 5533 | 5533 | 3000 |
| 5. Quantitative performance | 0000 | 2000 | 3242 | 0000 | 0000 | 3000 | 3242 | 0000 | 0000 | 4102 | 3342 | 0000 | 4102 | 3342 | 3342 | 0000 |
| 6. Operating range | 0000 | 2000 | 3242 | 0000 | 0000 | 3000 | 3242 | 0000 | 0000 | 4102 | 3342 | 0000 | 4102 | 3342 | 3342 | 0000 |
| 7. Safety | 2000 | 5000 | 4311 | 3000 | 2000 | 5000 | 4311 | 3000 | 2000 | 5314 | 4111 | 5102 | 5311 | 4111 | 4111 | 5102 |
| 8. Simulation—behavioral | 0000 | 1000 | 5334 | 0000 | 0000 | 1000 | 5334 | 0000 | 0000 | 4110 | 5334 | 0000 | 4111 | 5334 | 5334 | 0000 |
| **B. Functional description** | | | | | | | | | | | | | | | | |
| 9. Functional partitioning | 4424 | 5325 | 5322 | 3000 | 4424 | 5433 | 5433 | 3000 | 4424 | 5433 | 5433 | 3110 | 4424 | 5444 | 5544 | 2000 |
| 10. Form factor | 5555 | 2000 | 0000 | 3000 | 5555 | 3000 | 2000 | 3000 | 5555 | 4324 | 2000 | 5555 | 5555 | 5555 | 2000 | 5111 |
| 11. Algorithmic description | 0000 | 2223 | 5543 | 0000 | 0000 | 2223 | 5553 | 0000 | 0000 | 4324 | 5554 | 0000 | 4324 | 4324 | 5554 | 0000 |
| 12. Interface control and limit | 3553 | 2000 | 1000 | 3000 | 3553 | 3323 | 1000 | 3000 | 3553 | 3323 | 1000 | 5555 | 3553 | 4323 | 1000 | 5111 |
| 13. Environmental test parameters | 2000 | 2000 | 1000 | 3000 | 2000 | 3323 | 1000 | 3000 | 2000 | 3323 | 1000 | 5335 | 3323 | 3323 | 1000 | 5111 |
| 14. Simulation and functional | 0000 | 0000 | 5544 | 1000 | 0000 | 2111 | 5544 | 1000 | 0000 | 4432 | 5544 | 3110 | 5322 | 5554 | 5554 | 2000 |

## C. Logical description (digital)

| | | | | | | | | | | | | | | | |
|---|---|---|---|---|---|---|---|---|---|---|---|---|---|---|---|
| 15. Symbol definition | N/A | N/A | N/A | N/A | N/A | 3311 | 2000 | 5111 | 3311 | 5555 | 2000 | 5555 | 3311 | 2000 | 5555 |
| 16. Signal | N/A | N/A | N/A | N/A | N/A | 3000 | 5221 | 3000 | 3000 | 5555 | 5554 | 5422 | 3000 | 5554 | 5422 |
| 17. Ports | N/A | N/A | N/A | N/A | N/A | 2000 | 5221 | 3000 | 2000 | 5555 | 5554 | 5422 | 2000 | 5554 | 5422 |
| 18. Timing (description) | 4000 | 4000 | 4000 | 4000 | 5111 | 5332 | 4000 | 5555 | 4000 | 5555 | 5554 | 4000 | 4000 | 5554 | 4000 |
| 19. Simulation—logic (including models) | N/A | N/A | N/A | N/A | 4111 | 5332 | 4000 | 5324 | 1000 | 5555 | 5554 | 4000 | 1000 | 5554 | 4000 |

## D. Circuit definition (analog)

| | | | | | | | | | | | | | | | |
|---|---|---|---|---|---|---|---|---|---|---|---|---|---|---|---|
| 20. Symbols | N/A | N/A | N/A | N/A | N/A | 3311 | 2000 | 5555 | 3311 | 5555 | 2000 | 5555 | 3311 | 2000 | 5555 |
| 21. Gain charts/V–I plots | N/A | N/A | N/A | N/A | N/A | 5000 | 3000 | 5000 | 5000 | 5000 | 3000 | 5000 | 5000 | 3000 | 5000 |
| 22. Frequency plots | N/A | N/A | N/A | N/A | N/A | 5000 | 3000 | 5000 | 5000 | 5000 | 3000 | 5000 | 5000 | 3000 | 5000 |
| 23. Propagation delays | 1000 | 5010 | 3000 | 5010 | 5010 | 2000 | 5010 | 5555 | 3000 | 5555 | 5010 | 3000 | 3000 | 5010 | 3000 |
| 24. Timing description | 4000 | 4000 | 4000 | 4000 | 4000 | 4000 | 4000 | 5555 | 4000 | 5555 | 4000 | 4000 | 4000 | 4000 | 4000 |
| 25. Quantitative performance | 1000 | 3000 | 2000 | 5001 | 4112 | 2000 | 4112 | 5555 | 2000 | 5555 | 4112 | 3000 | 4112 | 4112 | 3000 |
| 26. Operating range | N/A | N/A | N/A | N/A | N/A | 3000 | N/A | 4332 | 3000 | 4332 | 2000 | 3000 | 3000 | 2000 | 3000 |
| 27. Q, R and M calculations | N/A | N/A | N/A | N/A | N/A | 2000 | N/A | 4000 | 2000 | 4000 | 1000 | 2000 | 2000 | 1000 | 3000 |
| 28. Simulation (circuit) | N/A | N/A | N/A | N/A | N/A | 1000 | 2110 | 5222 | 1000 | 5222 | 2110 | 1000 | 2110 | 2110 | 3000 |

Scoring criteria: For each of the four evaluation criteria, a score representing the suitability of each standard is assigned. The scoring criteria, are chosen to reflect similar levels of attainment for similar scores.

5 = Fully capable or substantially implemented, accepted and supported.

4 = Should be considered closer to a five than a three.

3 = Intended to convey a marginal condition: i.e., could be made to work with significant efforts.

0 – 2 = Signify that the standard is so far removed from applicability that it is probably not worth the effort to apply it to information item.

**Table 1.5. Arithmetic Design Process Functional Matrix [1]**

| DESIGN PROCESS | SYSTEM | | | | BOX | | | | PRINTED BOARD | | | | COMPONENT | | | |
|---|---|---|---|---|---|---|---|---|---|---|---|---|---|---|---|---|
| | IGES | EDIF | VHDL | IPC | IGES | EDIF | VHDL | IPC | IGES | EDIF | VHDL | IPC | IGES | EDIF | VHDL | IPC |
| **A. BEHAVIORAL DESCRIPTION** | | | | | | | | | | | | | | | | |
| 1. General | | ▨ | ⊠ | ⊠ | | ▨ | ⊠ | | | ■ | ■ | ▨ | ■ | ■ | | ■ |
| 2. Quality Level | | ▨ | ■ | ■ | | ▨ | ■ | | | ▨ | ■ | ▨ | ■ | ■ | | ■ |
| 3. Signal | | | ■ | ■ | | ▨ | ■ | | | ▨ | ■ | | ▨ | ■ | | |
| 4. Ports | | | ■ | ⊠ | | ▨ | ■ | | | ▨ | ⊠ | | ▨ | ⊠ | | |
| 5. Quantitative Performance | | | | | | | ■ | | | □ | ⊠ | | □ | ⊠ | | |
| 6. Operating Range | | ▨ | ⊠ | ⊠ | | ▨ | ⊠ | | | □ | ⊠ | | □ | | | ■ |
| 7. Safety | | | ■ | ■ | | | ■ | ■ | | ■ | ■ | ■ | ■ | ■ | | |
| 8. Simulation— Behavioral | | | | | | | | | | | | | | | | |
| **B. FUNCTIONAL DESCRIPTION** | | | | | | | | | | | | | | | | |
| 9. Functional Partitioning | □ | ■ | ▨ | | □ | ■ | ■ | ■ | □ | ■ | ■ | ■ | □ | ■ | ■ | ▨ |
| 10. Form Factor | | | | | | | | | ■ | | | | ■ | | | ■ |
| 11. Algorithmic Description | ⊠ | | | ■ | ⊠ | | ■ | | | | | ■ | | | ▨ | ■ |
| 12. Interface Control and Limit | □ | □ | □ | □ | | | | | □ | | | | □ | ⊠ | | |
| 13. Environmental Test Parameters | | | | | | ⊠ | ⊠ | | | ⊠ | | | ▨ | | | |
| 14. Simulation and Functional | | | ■ | ■ | | | | ■ | | ■ | ■ | ■ | | | ■ | ■ |
| **C. LOGICAL DESCRIPTION (DIGITAL)** | | | | | | | | | | | | | | | | |
| 15. Symbol Definition | -- | -- | -- | -- | | ■ | ■ | ■ | | ■ | ■ | ■ | | ■ | ■ | ■ |
| 16. Signal | -- | -- | -- | -- | | ■ | ■ | ■ | | ■ | ■ | ■ | | ■ | ■ | ■ |

24

| DESIGN PROCESS | SYSTEM | | | | BOX | | | | PRINTED BOARD | | | | COMPONENT | | | |
|---|---|---|---|---|---|---|---|---|---|---|---|---|---|---|---|---|
| | IGES | EDIF | VHDL | IPC | IGES | EDIF | VHDL | IPC | IGES | EDIF | VHDL | IPC | IGES | EDIF | VHDL | IPC |
| **C. LOGICAL DESCRIPTION (DIGITAL) (continued)** | | | | | | | | | | | | | | | | |
| 17. Ports | | | | | | | | | | | | | | | | |
| 18. Timing (Description) | | | | | | | | | | | | | | | | |
| 19. Simulation—Logic (including models) | | | | | | | | | | | | | | | | |
| **D. CIRCUIT DEFINITION (ANALOG)** | | | | | | | | | | | | | | | | |
| 20. Symbols | | | | | | | | | | | | | | | | |
| 21. Gain Charts/ V-1 Plots | | | | | | | | | | | | | | | | |
| 22. Frequency Plots | | | | | | | | | | | | | | | | |
| 23. Propagation Delays | | | | | | | | | | | | | | | | |
| 24. Timing Description | | | | | | | | | | | | | | | | |
| 25. Quantitative Performance | | | | | | | | | | | | | | | | |
| 26. Operating Range | | | | | | | | | | | | | | | | |
| 27. Q, R and M Calculations | | | | | | | | | | | | | | | | |
| 28. Simulation (Circuit) | | | | | | | | | | | | | | | | |
| **E. SIMULATION** | | | | | | | | | | | | | | | | |
| 29. Fault Simulation | | | | | | | | | | | | | | | | |
| 30. Test Vectors | | | | | | | | | | | | | | | | |
| 31. Thermal | | | | | | | | | | | | | | | | |
| 32. Vibration | | | | | | | | | | | | | | | | |

25

**Table 1.5.** *(Continued)*

| DESIGN PROCESS | SYSTEM | | | | BOX | | | | PRINTED BOARD | | | | COMPONENT | | | |
|---|---|---|---|---|---|---|---|---|---|---|---|---|---|---|---|---|
| | IGES | EDIF | VHDL | IPC | IGES | EDIF | VHDL | IPC | IGES | EDIF | VHDL | IPC | IGES | EDIF | VHDL | IPC |
| **F. NETLIST** | | | | | | | | | | | | | | | | |
| 33. Design Rules | | | | | | | | | | | | | | | | |
| 34. Parts | | | | | | | | | | | | | | | | |
| 35. Interconnectivity | | | | | | | | | | | | | | | | |
| **G. PHYSICAL DESIGN LAYOUT** | | | | | | | | | | | | | | | | |
| 36. Physical Design Rules | | | | | | | | | | | | | | | | |
| 37. Dimensions/Tolerances | | | | | | | | | | | | | | | | |
| 38. Package Interfaces | | | | | | | | | | | | | | | | |
| 39. Material Properties | | | | | | | | | | | | | | | | |
| 40. Reference Designators | | | | | | | | | | | | | | | | |
| 41. Cabling/Conductors | | | | | | | | | | | | | | | | |
| 42. Detailed Thermal Analysis | | | | | | | | | | | | | | | | |
| 43. Detailed R & M Analysis | | | | | | | | | | | | | | | | |
| **H. PHYSICAL DOCUMENTATION** | | | | | | | | | | | | | | | | |
| 44. Detail/Package Drawings | | | | | | | | | | | | | | | | |
| 45. Reference Designators | | | | | | | | | | | | | | | | |
| 46. Dimensions and Tolerances | | | | | | | | | | | | | | | | |
| 47. Material Construction | | | | | | | | | | | | | | | | |

26

| DESIGN PROCESS | SYSTEM | | | | BOX | | | | PRINTED BOARD | | | | COMPONENT | | | |
|---|---|---|---|---|---|---|---|---|---|---|---|---|---|---|---|---|
| | IGES | EDIF | VHDL | IPC | IGES | EDIF | VHDL | IPC | IGES | EDIF | VHDL | IPC | IGES | EDIF | VHDL | IPC |
| **H. PHYSICAL DOCUMENTATION** (continued) | | | | | | | | | | | | | | | | |
| 48. Assembly Drawings and Notes | ■ | ■ | ■ | — | ■ | ■ | ■ | □ | ■ | ▩ | ▩ | ■ | □ | ■ | ■ | ▩ |
| 49. Parts list | | | | | | | | | ▩ | ▩ | ▩ | | ■ | ■ | ■ | ■ |
| 50. Fixturing | | | | | | | | | | | | | ■ | ■ | ▩ | |
| 51. Pattern Geometry | — | | — | — | — | | — | — | ▩ | ▩ | ▩ | ▨ | ■ | ■ | ▩ | ■ |
| 52. NC Data | — | | — | — | ■ | | | — | ▩ | | | | ■ | | | |
| **J. ASSEMBLY AND TEST** | | | | | | | | | | | | | | | | |
| 53. Assembly Specification | □ | □ | □ | □ | □ | □ | □ | □ | ■ | | ■ | ■ | ■ | | ■ | |
| 54. Test/Burn-in Requirments | □ | □ | □ | □ | □ | | □ | □ | | | | | | | | |
| 55. Other Q, R and M Testing | □ | | □ | □ | □ | | □ | □ | | | | | | | | |
| **K. INSTALLATION** | | | | | | | | | | | | | | | | |
| 56. Drawings | ■ | | ■ | | ■ | | ■ | ▩ | ▩ | | ■ | ▩ | □ | | ▨ | |
| 57. Tech Manuals | | | | □ | | | | ▨ | | | | ▨ | | | ▨ | |
| 58. Shipping Container Drawings | | | | | | | | | | | | | | | | |

**Legend**

■ Arithmetic best. Potential must be 5. (Potential is doubled, others are added, if within 2 points it is a tie.) Total of 8 or more

▨ Arithmetic best. Potential is not 5.

▩ Not arithmetic best. Potential is 5.

□ Not arithmetic best. Potential is not 5, worth mentioning.

□ Not arithmetic best. Potential is not 5.

▨ Not enough points.

— Not applicable

manufacture and support. The benefits expected to be achieved by CALS include:

- Reduced acquisition and support costs
- Elimination of duplicative, manual, error-prone processes
- Improved reliability and maintainability of designs directly coupled to computer-aided design/engineering (CAD/CAE) processes and databases
- Improved responsiveness of the industrial base, i.e., to be able to rapidly increase production rates or sources of hardware based on digital product descriptions.

As a result of CALS initiatives, the various design languages have been rated according to their capability for handling electrical product data descriptions and recommendations have been made for how to implement and use these languages, Tables 1.4 and 1.5.

## References

1. "Electronic Packaging Handbook," IPC-PD-355, Institute for Interconnecting and Packaging Electronic Circuits, Lincolnwood, IL 60646.
2. "Automated Design Guidelines," IPC-D-390, Revision A, February 1988, Institute for Interconnecting and Packaging Electronic Circuits, Lincolnwood, IL 60646.
3. "Guide for Digital Descriptions of Printed Board and Phototool Usage per IPC-D-350," IPC-D-358, May 1990, Institute for Interconnecting and Packaging Electronic Circuits, Lincolnwood, IL 60646.
4. "Library Format Description for Printed Boards in Digital Form," ANSI/IPC-D-354, February 1987, Institute for Interconnecting and Packaging Electronic Circuits, Lincolnwood, IL 60646..
5. Dieter Bergman, Institute for Interconnecting and Packaging Electronic Circuits, "IPC-D-350: One Standard for PC Design," *Printed Circuit Design*, December 1987.

# 2

# Integrated Circuits

The semiconductor industry has progressed dramatically over the past decade. Advances in microprocessors and peripheral integrated circuits have been fast and furious. As a result, the electronic equipment design engineer has a virtual menu of architectural options, Figure 2.1, to satisfy his requirements. Thus, the options range from the use off-the-shelf standard functions to the development of integrated circuits (ICs) that are unique to their design requirements. [2, 2]

## 2.1 PROCESS TECHNOLOGY [3, 4, 5, 6]

A circuit designer can work with one of several integrated circuit transistor structures. The two most popular structures are the bipolar and metal-oxide semiconductor (MOS) technologies. These are beginning to receive competition from the newly-emerging gallium arsenide (GaAs) and bipolar-complementary MOS (BiCMOS) structures.

Gate switching speed, power dissipation, and functional density are the basic differences separating the integrated circuit technologies at the low-integration level, Figure 2.2. However, these are no longer valid in the very-large-scale integration (VLSI) regime. Thus, the choice of integrated circuit process technology, like the choice of architecture, has become far more complicated.

The old rules of thumb, e.g., bipolar ECL for high-power and low-density applications, CMOS technology for low-power and high-density devices, and GaAs for the highest speeds, do not always apply. The answers to the selection questions lay in understanding the available technologies from the new perspective of identifying the application areas that are most generally suited for

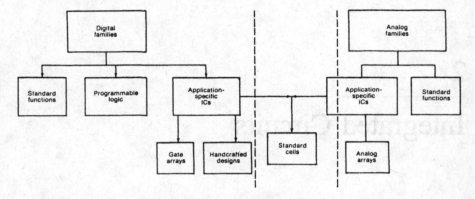

**FIGURE 2.1.**    Integrated circuit families.

each technology and the key system design trade-offs involved in the use of each technology.

### 2.1.1  Bipolar Technology

Bipolar transistors can be formed by either alloy or double-diffused techniques. However, all major bipolar integrated circuits use double-diffused transistors that are formed in an epitaxial (high-resistivity film) layer deposited on a silicon wafer. The epitaxial layer serves as the collector region of the transistor that allows for easy isolation of the separate circuit elements with the addition of a deep-diffused fence.

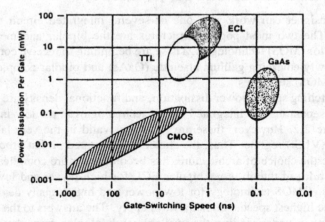

**FIGURE 2.2.**    Integrated circuit speed versus power.

Because of their formation process below the wafer surface, bipolar transistors are less subject to contamination. Less sensitivity to contamination is not the only advantage. They can also operate at voltages and current values that are compatible with other electrical devices. The switching speeds of bipolar transistors are also faster than those of the simpler MOS discrete transistors.

Using conventional bipolar structures, circuit speeds can be improved by wiring together the various elements into logic cells. Transistor-Transistor Logic (TTL) and Emitter-Coupled Logic (ECL) are the two most popular bipolar cell types.

TTL circuits have been popular with most system builders because of their combination of acceptable performance and ease of design. Alternately, ECL design has been a standard process only for computer builders who need high performance badly enough to put up with the difficulties created by the use of the higher-speed ECL circuits. For example, printed circuit board conductors must be designed with transmission-line, controlled-impedance techniques because the rise time of ECL signals puts them in the microwave frequency range.

## 2.1.2 Metal-Oxide Semiconductor (MOS) Technology

The formation of MOS integrated circuit devices is different from bipolar technology . The basic MOS fabrication process starts with an incoming wafer that is processed through oxidation and goes directly to masking. Thus it eliminates the epitaxial layer and isolation diffusion steps required in bipolar technology. The absence of the isolation structure allows for a higher MOS component density than is possible with bipolar technology.

In both bipolar and MOS technology the area of the circuit that does not contain active devices is called the field. A particular problem arises over the field in MOS circuits. Metal conductors running on top of the field oxide form a capacitor with the silicon below. If the voltage on the conductor becomes high enough, the field capacitor will create a charge in the underlying silicon, and thus cause shorted devices. Additional wafer processing is required to overcome this problem. They also require more careful handling during the printed board assembly process.

Unlike the simpler metal-gate processing, Complementary MOS (CMOS) requires more fabrication steps than are necessary in the bipolar process. However, this additional processing results in CMOS devices that have lower power consumption and increased speed that are attractive for many applications.

Bipolar technology is inherently more radiation resistant than most CMOS processes. Thus the advantages of designing with CMOS has resulted in additional processing steps that can be taken to enhance CMOS survivability.

Other CMOS processing variations prevail to customize particular features of the end product. This, in many cases, makes CMOS the most complicated and the most involved basic technology in the industry.

### 2.1.3　Bipolar-CMOS (BiCMOS) Technology

The adoption of a combined bipolar and CMOS technology by silicon foundries may well be the most important processing advancement in recent years. The ability to mix low-power, high-density CMOS circuit elements with faster, high-drive bipolar elements provides an optimal solution for many applications.

BiCMOS is implemented in two basic ways. One is to start with a CMOS process and add bipolar transistors to it. With CMOS as the underlying technology, users get good density but only a little increase in speed and performance. The other approach is to start a bipolar process and add CMOS transistors. With ECL as the underlying technology, the result is higher speeds but more limited density.

### 2.1.4　Gallium Arsenide (GaAs) Technology [7, 8]

Gallium arsenide (GaAs) technology is beginning to emerge from its limited niche as a high-priced, low-yield process with limited applications. Improvements in GaAs technology have resulted in higher yields of high-performance devices that are becoming price competitive with bipolar and CMOS process technologies.

Gallium arsenide (GaAs) is an ideal alternative to silicon as a semiconductor medium for achieving very-high speed electronic devices and integrated circuits. This is because its energy-band structure is such that electrons in GaAs are exceedingly "light" and highly mobile. Thus, electron velocities measured in GaAs structures range up to about five times those achieved in bipolar and MOS silicon-based devices. Furthermore, GaAs is readily available in a semi-insulating substrate form that substantially reduces parasitic capacitances, so that its outstanding device speeds can be more fully realized in integrated circuits, Table 2.1. This high speed, plus power dissipation that tends to be substantially lower than that of high-speed silicon devices, accounts for the growing interest in gallium arsenide.

Although a number of different GaAs integrated circuit families exist, it is

**TABLE 2.1.　Typical High-Speed Integrated Circuits**

|  | GaAs | ECL | CMOS |
|---|---|---|---|
| Critical dimensions ($\mu$m) | 1.0 | 2.0 | 2.0 |
| Chip size (mm$^2$) | 4.6 × 4.3 | 6.6 × 6.8 | 10 × 10 |
| No. of equivalent gates | 1200 | 2500 | 20-K |
| Time delay (ps/gate) | 375 | 510 | 1500 |
| Power dissipation ($\mu$W/gate) (50-MHz clock) | 190 | 1200 | 310 |
| Speed-power (fJ) | 71 | 610 | 465 |

possible to draw some general conclusions about the advantages of gallium arsenide technology over silicon-based technology.

- Currently, for the same power consumption, GaAs is about one-half an order of magnitude faster than Emitter-Coupled Logic (ECL), the fastest silicon-based family.
- GaAs is more radiation-hardened than is silicon. However, the difference is difficult to quantify.
- Also, GaAs is better suited to the efficient integration of electronic and optic components. If this is developed to the appropriate level, this combination may have a major impact on system design.

Basic disadvantages of GaAs technology as compared with silicon-based technology include the following.

- Gallium arsenide wafers can exhibit "dislocations" in the form of irregularities. Consequently GaAs die are generally smaller, have a lower transistor count, and have a poorer yield.
- GaAs substrates are more expensive than silicon substrates. Moreover, GaAs is brittle, such that wafers can be more easily damaged during the fabrication of integrated circuits.
- The noise margin of GaAs is generally not as good as that of silicon. Thus, it is often necessary to trade off die size against higher reliability.
- Lastly, some companies are having problems testing the very high-speed GaAs devices. But, this situation is expected to improve.

## 2.2 APPLICATION-SPECIFIC INTEGRATED CIRCUITS [2, 9, 10, 11]

Electronic system design is becoming more complex, with architectures reaching new heights of sophistication and expanding integrated circuit choices. Yesterday's systems were composed mostly of off-the-shelf small-scale (SSI) and medium-scale (MSI) levels of circuit integration. However, many of the new designs are being implemented with customized large-scale (LSI) and very-large-scale (VLSI) integration. Leading the migration toward LSI/VLSI are application-specific integrated circuits (ASICs) with functions tailored to the user's requirements.

### 2.2.1  ASIC Selection

Although many ASIC device families have overlapping attributes, some may depend merely on personal preference, many key attributes clearly separate the families. Thus, designers who have a knowledge of the available ASIC types

**TABLE 2.2.** ASIC Technologies

| Type of ASIC | | Architectural capabilities | Density (gates) | Development time | NRE charges | Production volumes | Production charges in $ per 1,000 gates |
|---|---|---|---|---|---|---|---|
| Full custom | | RAM ROM Analog | 1,000– 100,000 | Long | $50,000 | <2,500 2,500–10,000 >10,000 | N/A $2–$3 $1 |
| Standard and compiled cells | PC-based | RAM ROM Analog | 1,000– 10,000 | Moderate | $15,000 | <2,500 2,500–10,000 >10,000 | $5–$10 $3–$4 $2–$3 |
| | Work-station-based | RAM ROM Analog | 1,000– 50,000 | Moderate | $50,000 | <2,500 2,500–10,000 >10,000 | $5–$10 $3–$4 $2–$3 |
| Gate arrays | | Logic only | 1,000– 50,000 | Moderate | $15,000– $100,000 | <2,500 2,500–10,000 >10,000 | N/A $3–$4 $2–$3 |
| Field Programmable devices | PLDs | Fixed logic | 500 | Short | <$5,000 | <2,500 2,500–10,000 >10,000 | $8 $7 $6 |
| | FPGAs | Fixed logic | 1,000– 3,000 | Moderate | $5,000– $20,000 | <2,500 2,500–10,000 >10,000 | $10–$20 $7–$15 $5–$12 |

can make rational choices from among them. The choices include program-mable logic devices, gate arrays, standard and compiled cells, and full custom ASICs, Table 2.2. Among the key performance factors that separate the various devices are the speed at which they can operate, the power requirements of each, and the circuit density each device type affords. [12] Of course, the ASIC's procurement and nonrecurring engineering (NRE) costs, such as the cost of making masks, if any, and the design and development time are also important, Figure 2.3. Also, the determining factors in making the choice can be the design flexibility and packaging aspects of the device, the availability of software tools for computer-aided design (CAD), and adequate vendor product and application support.

The number of gates per device is generally the first ASIC feature that influences the selection process. For example, programmable logic devices (PLDs) have a relatively low 500 to 3,000 equivalent gates per chip. Gate arrays are next with up to 50,000 gates or more for specific applications. Standard and compiled cells have still higher gate counts. Fully-customized ASICs are the most complex, with up to 100,000 or more switching transistors per chip being common.

The time to prototype and develop an ASIC is often a major concern when making the choice as to which device is best suited for an application. Because PLDs are readily available off-the-shelf, prototypes can be obtained almost immediately at a relatively negligible development cost. Gate array prototype de-

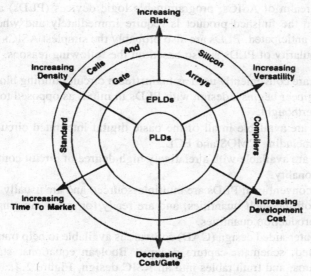

**FIGURE 2.3.**    ASIC technology tradeoffs.

velopment time is typically from 7 to 13 weeks, with prototype costs that are typically from $15,000 to $100,000.

Standard and compiled cells can take even longer, typically from 13 to 26 weeks, because their wafers are made from scratch instead of being fully or partially preprocessed, as are PLDs and gate arrays. Also, standard and compiled cell prototyping costs are usually between $15,000 and $50,000 or more. Fully-custom devices, obviously, have the longest prototyping times, often taking between 9 months to 2 years, with development costs usually ranging from $50,000 to over $100,000.

Thus, programmable logic devices are often the logical choice when the product is required immediately or when frequent design changes are likely to occur. PLDs are limited though to relatively simple requirements when compared to the functionality and densities of the other ASIC technologies.

If functional and quantity requirements are high, the use of gate arrays is often the most feasible, with standard and compiled cell devices the choice for the higher quantity demands. However, the better silicon usage characteristics of standard cells may offset their higher development costs if the number of services being used is suitable. The selection of a fully-custom ASIC is generally acceptable when the largest production quantities are involved and if the long turnaround time can be tolerated.

### 2.2.2  Programmable Logic Devices [13, 14]

Within the realm of ASICs, programmable logic devices (PLDs) are the best choice when the finished product is require immediately and when frequent changes are anticipated. PLDs are also probably the simplest ASICs to use. The general popularity of PLDs is also based on the following reasons.

- PLDs are conveniently used as mainstream circuit building blocks, so that an engineer begins a design with PLDs in mind, as opposed to their being an afterthought.
- PLDs are available in all of the basic digital integrated circuit technologies, including CMOS and ECL.
- PLDs are available with a relatively high degree of circuit complexity and functionality.
- Most conventional PLDs are multiple-sourced and are usually available in large off-the-shelf quantities, and are ready for quick ramping up to volume production quantities.
- Computer-aided design (CAD) software is available to help transform user-specified, schematic-capture designs, Boolean equations, state-machine diagrams, and truth tables into an ASIC design, Figure 2.4.

Conventional PLDs with their fuses can be programmed only once. However, Erasable PLDs (EPLDs) are changeable and 100% testable before pro-

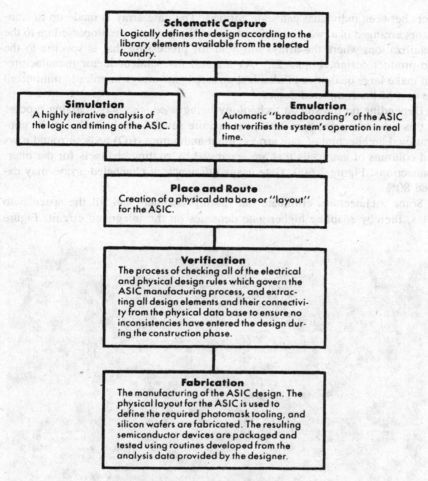

**FIGURE 2.4.** ASIC design process.

gramming. This affords the designer a level of versatility during the logic design process.

In certain applications EPLDs match the densities of gate arrays, without the associated tooling costs, extended development schedules, and reduction of in-house design control. Thus, EPLDs fit between PLDs and gate arrays in terms of complexity, versatility, and cost per function.

### 2.2.3  Gate Arrays [1, 6, 15]

Unlike standard cells, gate arrays are actually standard devices that become application specific in the final stages of wafer processing, where the interconnects between individual gates are formed. The gate array is made up of transistors arranged in a two-dimensional array that has been preprocessed up to the metallization, where the arrays are held for processing that is specific to the end-product design, Figure 2.5. As a result the semiconductor manufacturer can make large quantities of identical unwired gate arrays in order to minimized the cost of the completed device.

Depending on the silicon technology being used, the personalizing process of this type of semicustom ASIC can require several mask levels. In the conventional architecture of gate arrays, input and output (I/O) pads surround rows and columns of gate cells that are separated by routing channels for the interconnections, Figure 2.6(a). Gate usage efficiency of channeled arrays may exceed 80%.

Some architectures eliminate the routing channels and fill the space with gates, thereby enabling higher gate densities on the integrated circuit, Figure

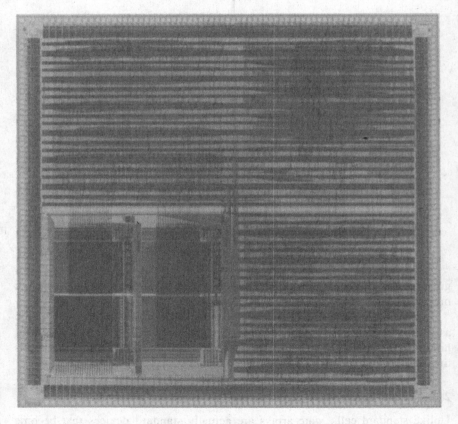

**FIGURE 2.5.**    20,000-gate CMOS Array (*Fujitsu Microelectronics*).

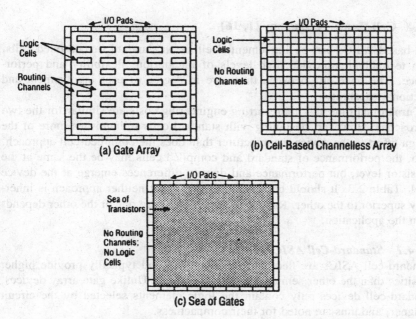

**FIGURE 2.6.**    Gate-array architectures.

2.6(b). This architecture puts the interconnections on a separate metallization layer.

Another approach, called sea of gates, does not use gate cells, but rather a "sea" of uncommitted transistors. This provides the possibility of having even higher circuit densities, Figure 2.6(c). However, the efficient utilization of gates decreases from 35% to 50% because transistors are lost, unless triple-level metallization is used.

Although gate arrays are not as flexible to use as are standard-cell ASICs, they are easier to implement, and design costs are typically less, and development times typically shorter. Gate arrays also come in a wide variety of sizes, but the degree of complexity they can handle is fixed. Too large an application just will not fit. Conversely, with too small an application, the design will be inefficient. Since the gate array's die size is fixed, the number of I/O terminals is also fixed.

Field programmable gate arrays (FPGAs) have emerged as a technology that offers many of the advantages of PLDs with greater density capabilities. Although the densities offered by FPGAs exceed those of PLDs, they are still considered to be lacking in complexity. [12] Their development costs are higher than PLDs, since the components themselves, the hardware programmer, and the computer-aided engineering (CAE) software needed to program them cost more. They are also more complex to program.

### 2.2.4 Cell-Based Technology [1, 16]

Cell-based ASICs can be implemented either as standard or compiled cells. Both technologies provide high levels of integration, flexibility and performance. The trade-offs between them are in degree of user involvement and functional capability.

Turnaround time and nonrecurring engineering costs are similar for the two approaches. However, designing with standard cells relinquishes more of the design effort to the device manufacturer than does the compiled-cell approach. Also, the performance of standard and compiled cells may be the same at the transistor level, but performance and density differences emerge at the device level, Table 2.3. It should be noted, however, that neither approach is inherently superior to the other. Rather, the decision to use one or the other depends upon the application.

#### 2.2.4.1  Standard-Cell ASICs

Standard-cell ASICs are flexible to design with and typically provide higher densities than the other semicustom ASIC options. Unlike gate array devices, standard-cell devices only contain the logic elements selected by the circuit designer, and thus are noted for their compactness.

Die size, a major factor in determining the cost of the device, is often the key factor in selecting standard-cell implementations, most of which is CMOS technology. Unfortunately, the number of I/O terminals required by the design often offsets any die-size advantage that standard-cell ASICs may have over other methodologies.

A standard-cell design can provide a degree of complexity that is proportional to the application. Thus, a design of any size can be 100% efficient. Accordingly, the standard-cell device's overall package type and cost can be proportional to the circuit complexity as opposed to being fixed, as is the case with gate arrays.

**TABLE 2.3.    Typical ASIC Circuit Implementation Alternatives**

|  | Cell compiler | Standard cell | Gate array |
|---|---|---|---|
| Core size | 1085 × 1112 | 1872 × 1740 | 2160 × 2112 |
| Number of transistors in core | 980 | 1286 | 1286 |
| Transistor density (Sq. mils per transistor) | 1.91 | 3.93 | 5.49 |
| Design time | 4 Days | 7 Days | 7 Days |
| Performance (Critical path delay per gate) | 45 ns | 68 ns | 49 ns* |

*The gate array design is implemented in a faster CMOS process technology.

A standard cell uses a library of simple, fixed cells and soft macrocells, i.e., predefined combinations of cells. User involvement typically includes logic partitioning, schematic entry, simulation and net-list verification. Fixed-height cells are then arranged by automatic placement and routing programs in fixed-height rows that are separated by variable-width routing channels.

Standard-cell placement and routing is usually handled by the device manufacturer, rather than by the user. After the place-and-route operation, the design is resimulated using actual wiring delay figures to help ensure accurate performance estimation.

The most limiting factors on using standard-cell devices are the relatively-long development period and comparatively-high nonrecurring engineering (NRE) costs. Unlike gate arrays, that are personalized in the last few steps of wafer processing, standard cells are totally unique to the device. Thus, none of the manufacturing steps can be shared by multiple applications.

### 2.2.4.2  Compiled-Cell ASICs

A cell compiler is a design tool for automatically generating the device layout, simulation models, and schematic symbols that are targeted for a specific function. However, since one compiler cannot generate layouts for all functions, a design environment usually includes a library of cell compilers or parameter-based software modules.

In concept, an engineer can design a device by identifying compiler functions that correspond to the block diagram description of the circuit being implemented. These functions are subsequently compiled by the software. The design process then proceeds in a manner that is similar to the design process used for standard-cell devices.

## 2.2.5  Fully-Custom Devices [12]

Fully-custom ASICs are the most complex type of devices. They are an all-level technology, which means that all manufacturing tooling (photomasks) are customized.

Designing a fully-custom ASIC is similar to programming in assembly language. Although the end circuit performance results are optimized, the design time is long and must be accounted for in the development of the end-product equipment.

An advantage of using fully-custom ASICs, as opposed to using semicustom devices, is that the highest speed and/or greatest circuit density is possible because the design is handcrafted with individually-tuned transistors, rather than being implemented with an automated process.

These performance and complexity benefits, however, come at some cost. The design and verification tools are very expensive and complex to use. Thus, because of the time and expenditures involved, fully-custom ASICs are generally only used in large-volume applications.

## 2.2.6   Analog and Mixed ASICs [3]

ASICs are not limited to digital technology. They are also available in pure analog form or in mixed analog and digital designs. Linear arrays offer a predefined tile arrangement of generic analog functions fabricated by the silicon foundry. Metallization interconnects the diodes, resistors, transistors, capacitors, and other devices to form more complex analog functions, in much the same manner that it interconnects the core logic elements of a digital gate array. The difference is that instead of having one homogeneous core logic element to deal with, the analog designer must use the limited number of elements in the tile array.

This constraint does not exist with analog standard cells. With these cells, the designer simply selects cells that have the needed characteristics, places them, and routes them to build a new ASIC. Still, the added difficulties of compiling an analog function have hindered the development of analog cell compilers.

Mixed analog and digital ASICs exist where both types of functionality reside on a single semicustom die. In such cases, the gate-array approach is not feasible, since linear functions implemented with gates offer very limited resolution and accuracy.

## References
1. Bill Chow, Integrated CMOS Systems Inc., "The ASIC Phenomenon," *Circuit Design*, January 1990, pp. 39–44.
2. Dave Bursky, Editor, "Semicustom ICs - 1984 Technology Forecast," reprinted with permission from *Electronic Design*, Vol. 32, No. 1, January 12, 1984, pp. 206–228.
3. Jon Gabay, Editor, "ASIC Technologies Boosting Chip Density and Speed," *High Performance Systems*, May 1990, pp. 32–53.
4. Peter VanZant, Semiconductor Services Inc., "Integrated Circuit Fabrication," *Microelectronics Manufacturing and Testing*, October 1985, pp. 16–18.
5. Warren Andrews, Editor, "Options Grow as Programmable Devices Join Standard Parts and Semicustom Circuits," *Electronic Engineering Times*, September 29, 1986, pp. T4, T43, T46.
6. Larry Jack, David Wick, Bill Woodruff, Honeywell Inc., "GaAs or Silicon: Which Should You Use?," *Electronic Engineering Times*, July 14, 1986, pp. T54–T56.
7. Chappell Brown, Editor, "ECL and GaAs Move into the Commercial Mainstream," *Electronic Engineering Times*, July 21, 1986, pp. 37, 44.
8. Thomas M. Reeder, Ajit G. Rode, Tektronix Inc., "High Speed Systems Look to GaAs for Lower Power LSI," *Computer Design*, September 1984.

9. Ronald Collet, Senior Technical Editor, "ASICS Take Your Pick," *Digital Design*, June 1986, pp. 29–36.
10. Dev Chakravarty, Motorola Inc., "Semicustom ASICs Solve Costly Problem of Lengthy Design Times," *Electronic Products*, March 1, 1985.
11. Keith Gutierrez, Texas Instruments Inc., "Which ASIC is Best for You?" *Circuit Design*, January 1990, pp. 12–22. Reprinted with permission of Texas Instruments.
12. Jackie Batty, IC Designs Inc., "The Basics of ASICs," *Printed Circuit Design*, August 1989, pp. 38–44.
13. Joseph Vithayathil, National Semiconductor Corp., "PLDs are Now a Real Option for ASIC Users," *Electronic Engineering Times*, September 29, 1986, pp. T14–T18.
14. David A. Laws, Altera Corp., "EPLDs as an Alternative to Gate Arrays for Custom Logic Design," *Electronic Engineering Times*, May 12, 1986, pp. T28, T73.
15. Dave Bursky, Editor, "Subnanosecond Silicon ECL Gate Arrays Face Challenge from GaAs and CMOS," Reprinted with permission from *Electronic Design*, Vol. 32, No. 6, June 12, 1986, pp. 74–84.
16. Suresh Dholakia, Antonio Martinez, Russell Steinweg, VLSI Technology Inc., "The Relative Merits of Standard and Compiled Cells," *Electronic Engineering Times*, May 12, 1986, pp. T18, T24.

# 3

---

# Circuit Component Packages

The variety of package types, materials, and lead counts available for electronic equipment components is quite extensive. As a result, the careful selection of the right package is a primary concern for both end product equipment and component manufacturers.

The consequences of making the wrong choice can be considerable. It can mean not only that more will have to be paid for the component, but also for the processing needed to assemble it. [1] Thus, all components should be qualified for the assembly process to be used. The physical dimensions of the component should be compatible with the assembly handling equipment. Also, the parts must not be excessively degraded physically or electrically by the assembly and end product environments to which it will be exposed.

## 3.1 ELECTRONIC CIRCUIT COMPONENTS [2, 3]

Electronic circuit components in electronic component packages serve to protect the devices within them from the environment, provide communication links with other components, remove heat, and provide a means for handling and testing. In large, high density integrated circuits (ICs), these functions are not only important, they are also challenge to meet.

There is an emphasis on maintaining or decreasing the procurement/assembled costs of the more complex packages. Also, the material composition, finish, and configuration of both the component package body and its terminations must be considered in the choice of the assembly method. This can be achieved by increases in functional density and performance by the use of enhanced memory, data processing, and application specific integrated circuits (ASICs).

There are several interrelated electronic component package design factors, some of which strongly influence the others. In the package design stage, one such factor may be emphasized over another for a given application. Examples of this might be emphasizing package electrical performance over cost for a high speed device, or emphasizing cost over performance for a high quantity application. Therefore, the characteristics of each package must be clearly understood in order to make the optimum selection.

In addition to the component package cost itself, it is important to consider the sum of the direct and indirect cost increments associated with component procurement (multiple sourcing), handling, assembly, testing, repair/rework, inventory, and reliability criteria. In other words, the electronic component packaging cost factor must be reviewed, not only in terms of related design trade-offs, but also in terms of the effect on overall end product cost.

### 3.1.1   Integrated Circuit Packages [1, 4]

For many years the dual inline package (DIP) predominated for use with integrated circuit (IC) components. However, the steady increase in IC package input/output (I/O) terminal count requirements and the emergence of surface mount technology (SMT) resulted in the use of a wide variety of IC component packages. Integrated circuit packages come in the form of leadless and leaded chip carriers, flat packs, and small-outline ICs for surface mounting applications, and pin-grid arrays for through-hole mounting, Table 3.1. [5]

Besides taking up far less space on printed circuit boards, than does the comparable DIP, the new packages can accommodate larger IC die and significantly more I/O terminals (and reduced terminal pitch) with improved electrical performance. However, because of the variety of packages available and the wide range of their costs, each type of IC package has its own advantages and limitations.

Thus, the selection process becomes a trade-off that takes into account many factors, including:

- Through-hole versus surface mounting
- Input/out terminal capacity (and pitch)
- Ceramic (hermetic) versus plastic dielectric
- Leaded versus leadless terminations
- Electrical performance
- Environmental stability
- Thermal management features
- Procurement and assembly costs.

**TABLE 3.1.  Characteristics of Common Integrated Circuit Packages [5]**

| Package type | Range of physical dimensions | Electrical characteristics[1] | Thermal characteristics[1] (°C/W) | Usable gates[3] | Relative cost (per pin) |
|---|---|---|---|---|---|
| Through-hole DIP | Number of pins: 16 to 64<br>Pin pitch: 100 mils<br>Body length: 75 to 2.3 in.<br>Body width: 0.300 to 0.700 in. | R: Medium<br>L: High<br>C: Low | Ceramic/plastic $\theta JA^2$:<br>70–40/<br>120–80 | Up to 17,000 gates | 1 |
| Surface mount SOIC | Number of pins: 16 to 28<br>Pin pitch: 10 mils<br>Body length: 50 to 70 mils<br>Body width: 0.300 to 0.400 in. | R: Medium<br>L: Medium<br>C: Low | Ceramic/plastic $\theta JA^2$:<br>110–80/<br>130–105 | Up to 6,500 gates | 6–Ceramic<br>2.5–Plastic |
| Surface mount OFPT | Number of pins: 48 to 260<br>Pin pitch: 10 mils<br>Body width: 0.65 to 1.7 in. | R: Medium<br>L: Medium<br>C: Low | Plastic $\theta JA^2$: 95–60 | Up to 17,000 gates | 6 |
| Surface mount CLCC | Number of pins: 28 to 84<br>Pin pitch: 40 to 50 mils<br>Body width: 0.45 to 0.97 in. | R: Medium<br>L: Medium<br>C: Medium | Ceramic $\theta JA^2$: 70–45 | Up to 25,000 gates | 30 |
| Surface mount PLCC | Number of pins: 28 to 84<br>Pin pitch: 50 mils<br>Body width: 0.49 to 1.19 in. | R: Medium<br>L: Medium<br>C: Low | Plastic $\theta JA^2$: 65–50 | Up to 17,000 gates | 2 |
| Through-hole PGA | Number of pins: 64 to 299<br>Pin pitch: 100 mils, 70 mils<br>Body width: 1,033 to 1.7 in. | R: Low/low<br>L: Low/low<br>C: High/low | Ceramic/plastic $\theta JA^2$:<br>40–19/<br>46–38 | Up to 75,000 gates | 60–Ceramic<br>12–Plastic |

[1] R = Resistance, L = Inductance, C = Capacitance
[2] Assuming static airflow.
[3] Assuming 1.5-μm CMOS technology.

46

### 3.1.1.1 Dual-Inline Packages (DIPs)

The dual-inline packages (DIPs) are available in both relatively low cost plastic, Figure 3.1 (bottom), and hermetic ceramic types, with through-hole mounting I/O terminals on a 2.54mm (0.100 inch) inline pitch. Their inline-to-inline terminal spacing varies from 7.62mm to 15.24mm (0.300 to 0.600 inch) depending on the I/O count.

Several factors have increased the popularity of the DIP for integrated circuits and other printed-board components, e.g., switches. Among these are:

- The DIP has reliably and cost-effectively satisfied the electrical and mechanical requirements for many years.
- The DIP's terminal count range of from 8 through 40 leads satisfies the needs of a wide range of IC devices.
- DIPs, as with all through-hole mounted components, can be readily soldered to printed boards manually or using mass termination techniques, such as wavesoldering.

**FIGURE 3.1.** Integrated circuit packages—dual-inline packages (bottom), leadless type a chip carrier (top left), cavity-down pin-grid array (top center), J-lead plastic chip carrier (top right). (*Courtesy of Siemens Corp.*)

- A wide variety of sockets are available for solderless DIP mounting to facilitate testing and maintenance.
- Intensive capital investments have been made in all types of DIP fabrication, handling, assembly, and testing equipment.

However, all of these advantages can be overshadowed by the need for more I/Os, relatively smaller packages, larger die cavities, and improved electrical performance by the ASICs and other major circuit devices. Although there are only a few of these enhanced IC devices in each circuit, they control the performance of the assembly. As such, their needs cannot be compromised to any great extent. Thus, the use of DIPs often cannot be justified in these applications.

### 3.1.1.2  Chip Carriers
Chip carriers can be generally described as being low profile, rectangular (usually square), surface mount integrated circuit packages with I/O terminals on all four sides. The I/O terminals consist of either metallized features on leadless versions or discrete leads formed around or attached to the side of the package on the leaded versions. The leadless chip carrier usually has a ceramic body, while the leaded types are usually plastic.

The Joint Electronic Device Engineering Council (JEDEC) established configurations for the original chip carriers that allowed for multiple design approaches, manufacturing techniques, and attachment means in order to allow for the choice of package that is best tailored for the application. JEDEC standardized both leadless and leaded chip carriers in two basic styles, one with 1.27mm (0.050 inch) terminal centers, and the other with 1.0mm (0.040 inch) spacing, and provided interchangeability within similar terminal spacing outlines.

The original 50-mil center family contained six square-package variations with up to 156 I/O terminals. Four leadless versions, types A, Figure 3.1 (top left), B, C and D, mount in different orientations depending on the type, mounting substrate, and preferred thermal orientation. (A rectangular Type E was subsequently added for use primarily with memory devices with up to 32 I/O terminals.) These packages are ceramic, with hermetically sealed metal or ceramic lids.

There are two leaded versions, a Type A that is a premolded or postmolded J-lead plastic package, Figure 3.1 (top right). The other version, Type B, is the leadless Type A package with clip-type leads attached to it to facilitate soldering to the interconnecting substrate.

### 3.1.1.3  Flatpacks
Flatpacks are among the oldest types of integrated circuit packages for surface mount applications. The typical dual row flatpack has up to 50 flat ribbon leads

that come straight out from its body on 1.27mm (0.050 inch) centers. Thus, the leads have to be formed downward, usually in a gullwing fashion, to facilitate surface mount assembly.

A newer version is the plastic quad flatpack (PQFP). The JEDEC-approved PQFP, Figure 3.2, is a high density package that has up to 244 gullwing leads that are preformed by the IC manufacturer on 0.63mm (0.025 inch) centers. The PQFP also features molded bumpers in its corners to help protect the leads from damage during handling and assembly.

Other four-row, plastic flatpacks with higher I/O counts and smaller terminal pitch are also available. One version uses a relatively low cost, easy to test, straight lead, tape automated bonding (TAB) construction. As supplied by the integrated circuit manufacturer, the tested/molded die is ready for excising from its own carrier frame and lead forming by the printed board assembler. Such devices are projected to be used with integrated circuits with from 40 to 350 or more leads.

### 3.1.1.4 Grid Arrays
The higher the I/O terminal count becomes, the lower the percentage of the total package area that any given die cavity will occupy. For optimum packaging density, this percentage should be as high as possible. Therefore, for high-I/O applications, fine-pitch (0.6 mm/0.025 inch centers or less) chip carriers/quad-flatpacks and grid array packages are often used.

Pin-grid array (PGA) packages are commonly used for through-hole mounting applications with one surface of the package populated with contacts usually on a 2.54 mm (0.100 inch) grid. The use of a solid grid of pins necessitates

**FIGURE 3.2.**    Plastic quad gullwing flat pack. (*Courtesy of Intel Corp.*)

placing the die cavity on the side of the package that is opposite to that with the contacts. The use of only a double-, Figure 3.1 (top center), or triple-row grid of pins permits having the die cavity and I/O contacts on the same surface of the package, thus allowing for the use of an optional thermal management heatsink on the other surface.

As can be seen in Figure 3.3, for I/O terminal counts in excess of 100, a 2.54 mm (0.100 inch) solid grid PGA provides greater I/O density than either the 1.27 mm (0.050 inch) or 1.0 mm (0.040 inch) chip carriers; however, the fine-pitch chip carriers (and quad-flatpacks) are clearly more I/O efficient.

Pin-grid array packages are usually ceramic, although some low cost plastic/glass-epoxy versions are being made. Also, fine-pitch (less than 100 mil) pin-grid array packages and pad-grid array packages, suitable for surface mount applications, and are being considered for high density applications, although they are not readily available and they require special assembly handling.

### 3.1.1.5  Small-Outline Integrated Circuits (SOICs)

With all of the recognition being given to surface mount chip carriers and flat-packs, small-outline integrated circuits (SOICs) are quite often used for low lead-count applications, i.e., devices with up to 28 leads. In appearance, SOICs

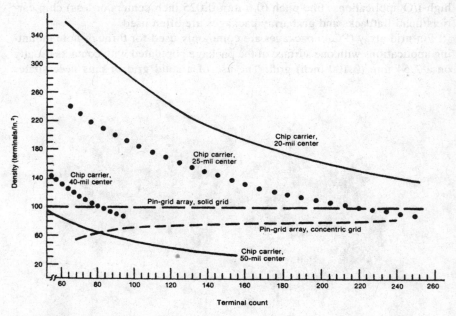

**FIGURE 3.3.**   Terminal count versus terminal density for various integrated circuit packages. [4]

**TABLE 3.2.    Small-Outline Integrated Circuit and Dual-Inline Package Comparison [2]**

| Item | 8-Pin SO | | 8-Pin DIP | 16-Pin SO | | 16-Pin DIP | 28-Pin SO | | 28-Pin DIP |
|---|---|---|---|---|---|---|---|---|---|
| Body size, L × W, mm$^2$ | 20 | (3.5) | 70 | 40 | (3.5) | 140 | 140 | (3.5) | 500 |
| Board area, mm$^2$ | 31 | (2.6) | 80 | 62 | (2.8) | 175 | 192 | (3.0) | 590 |
| Body thickness, mm | 1.45 | | 3.1 | 1.45 | | 3.6 | 2.45 | | 3.9 |
| Height above board, mm | 1.75 | | 4.2 | 1.75 | | 5.1 | 2.65 | | 5.1 |
| Weight, mg | 60 | (10) | 600 | 130 | (9) | 1200 | 700 | (6) | 4300 |

Ratios are shown in parentheses ( ).

resemble miniature versions of molded plastic DIPs with either gullwing or J-lead terminals.

The primary advantage of the SOIC is its small size and its suitability for surface mounting. As compared to the DIP, as shown in Table 3.2, the package is about one-third the size and uses about one-quarter of the substrate mounting area of the equivalent I/O DIP; its body thickness, height, and weight are also proportionately less. (Small-outline transistors, SOTs, that compare in a similar manner to conventionally packaged transistors, Figure 3.4 (bottom) are also available.) Consequently, small outline components are quite often used in applications where space and weight are at a premium.

### 3.1.1.6  "Carrier Frame" Packages

By the use of inert-molding technology, National Semiconductor Corp. took the previously untestable single-layer TAB tape construction (see Chapter 4) and crafted a low cost, easy-to-test integrated circuit package. As supplied by the integrated-circuit manufacturer, the tested/molded die in its own carrier frame packaging, with a guard ring, is ready for excising, lead forming and surface-mount soldering by the printed board assembler.

The package, called Tape-Pak®, is being used by several integrated circuit manufacturers to provide fine pitch packages with lead counts ranging from 20 to over 360 in one of four different standardized body sizes. Typical sizes are 7 mm (0.286 inch) square for 40-lead devices, and 20 mm (0.805 inch) square for the 132-lead configuration.

### 3.1.1.7  Transistor Outline "TO" Cans

The packaging technology for the early transistors was often in the form of three-lead, hermetically sealed, metal packages, that were referred to as tran-

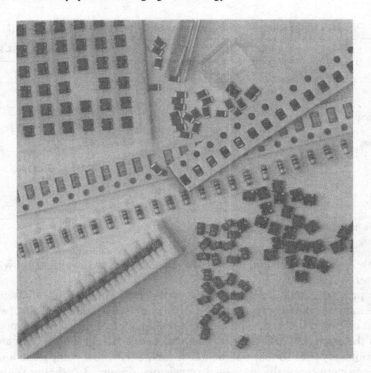

**FIGURE 3.4.**    Discrete electronic component packages—small-outline transistor (bottom), leadless chip components (top). (*Courtesy of Rohm Corp.*)

sistor-outline or "TO" cans, with the leads emanating radially outward from the base of the component package. This configuration was subsequently expanded for the packaging of integrated circuits with up to twelve leads, Figure 3.5.

### 3.1.2  Discrete Electronic Circuit Components [2, 6]

Electronic circuit components vary both in type and shape. In general, these components are usually selected for electrical, thermal, and/or mechanical characteristics that are determined by the requirements of the end product. Often the selection of such components is also dependent on the invoked specifications, availability, and/or cost.

The material composition, finish and configuration of both the body and the component's I/O terminals must be considered in the choice of the circuit assembly methods. Thus, many discrete electronic components, such as resistors and capacitors, are available in several configurations in order to facilitate their use in both through-hole and surface mount applications.

**FIGURE 3.5.**    Transistor outline "TO" can radial-lead integrated circuit. (*Courtesy of Burr-Brown Corp.*)

### 3.1.2.1  Discrete Leaded Components

Discrete circuit components with leads for through-hole mounting are of two basic types, axial-lead and radial-lead. Axial-leaded components with two leads are perhaps the most common type of electronic components used in printed board assemblies.

These components usually have cylindrical bodies with round wire leads that, as the name implies, exit from each of the opposite ends of the package along its major central axis. Many resistors, capacitors, and diodes are supplied in this configuration.

Discrete radial-lead components are also used in many printed circuit board assemblies. This type of component has all of its leads, usually two or three, exiting from only one end of the package. The actual body shape and construction is variable, some radial-lead components are basically round discs, others are rectangular in cross-section, and still others are in the form of transistor outline cans, Figure 3.5, with or without preformed leads or other component mounting provisions. Certain types of capacitors, diodes, transistors and variable resistors are often packaged as radial-lead components.

### 3.1.2.2  Discrete Leadless Components [6, 7]

Discrete leadless chip components suitable for surface mounting are basically miniature axial-lead components with the leads replaced by metallized terminals. They are available in both flat rectangular and cylindrical shapes, Figure 3.4 (top) for packaging resistors, capacitors, diodes and transistors.

Chip resistors are available in either a solderable wraparound or wire bonding termination configuration. The wraparound end-metallization resistor is terminated "face-up" so that post-attachment inspection and/or trimming is practical. More importantly, the resistive element itself is away from the solder joint and land pattern.

The wire-bonding type of leadless resistor has all of the termination metallization on the top of the chip. A chip-attach adhesive is usually used to hold the component in place as part of the assembly operation.

The most popular leadless chip capacitors have ceramic bodies, although silicon, porcelain, tantalum, and glass capacitors are also available. However, the ceramic types are used more extensively because of their ruggedness, ease of handling, wide range of values, high volumetric efficiency, and relatively low cost.

## 3.2 ELECTROMECHANICAL CIRCUIT COMPONENTS

A typical electronic system contains a mixture of electromechanical components, i.e., connectors, sockets, switches, etc. Associated with each of these components is discrete wiring, printed wiring, backplane wiring, and flat/round multiconductor cables. The electromechanical interface to this variety of wiring and cabling can be in the form of soldered, solderless, semipermanent, or permanent terminations. [8]

For optimum electronic packaging results, there must be a clear understanding of the components which are being interconnected. The electrical and mechanical requirements of the system usually determine the types of connectors and wiring/cable that can be used most cost effectively. However, there are many cases in which the wiring type and size has a direct effect on the connector selection and method of termination, i.e., coaxial cabling and fiber optics. Conversely, there are many considerations that are common to all electronic systems, Table 3.3.

### 3.2.1   Printed Board Connectors

Printed board connectors form a major part of the family of electronic connectors. They are specifically designed for use as interconnection devices between printed wiring/circuit boards or between boards and discrete wiring.

Printed board connectors can be divided into one-part (edge-board) and two-part (plug-and-receptacle) categories. Each category can be further subdivided into various styles or designs that vary mainly in contact configuration, the manner in which the contacts are retained, contact spacing, board thickness, and the manner in which the contacts are terminated.

**TABLE 3.3.    Electromechanical Component Considerations [8]**

| Parameter | Consideration |
|---|---|
| Frequency of mating and unmating | If very few mating cycles are involved, low cost, high pressure connectors can be used. |
| Insertion and withdrawal forces | Connectors requiring 10 lb of force can be engaged and disengaged readily by hand. Mechanical aids may be required for higher forces. |
| Number of contacts and spacing | Number of contacts affects insertion and withdrawal forces; spacing affects production speed. |
| Operating temperatures | Operating temperatures can dictate selection of materials (both metal and plastic) in connectors. |
| Humidity conditions | Corrosion (especially galvanic) is affected by humidity; moisture absorption can also change electrical properties of the insulation. |
| Barometric pressure | At high altitudes, outgassing of plastic contaminates sealed compartments. Voltage breakdown may be a limitation. |
| Atmospheric contaminants | The major contaminants present can determine metals, plastics and plating required. |
| Shock and vibration conditions | The plane and amplitude of expected shock or vibration affect connector design. A resonantly vibrating member can destroy electrical continuity. |
| Storage conditions | Will warehousing or nonoperating environment be the same as operating conditions? Shipping methods and conditions must be considered. |
| Current rating | Current capacity not only influences contact size and style, but minimum anticipated current in an active circuit can influence plating selection. |
| Contact resistance | Do not forget total number of connections in the largest path when specifying acceptable resistance. |
| Interconnect capacitance | Spacing, materials, and design are influenced by this parameter. |
| Keying requirements | To prevent mismating similar connectors, keying facilities may be needed. |
| Color coding and circuit identification | In some cases color coding or circuit identification may be needed instead of (or in conjunction with) keying and polarization. |
| Mounting methods | Are connection devices to be mounted from the front panel, with rivets, screws, friction devices or clasps, or must they be self-mounting (snap in)? |
| Maintenance requirements | Maintenance needs can influence thermination methods, mounting, tolerable mating forces, etc. |
| Projected storage time | Excessively long storage time permits films to form on some metal surfaces. Some insulating plastics can be adversely affected as well. Naturally, storage conditions are a factor here. |
| Life expectancy of end product | During design or after completion be careful of underspecifying. |

**TABLE 3.3.** *(Continued)*

| Parameter | Consideration |
| --- | --- |
| Approvals required (UL, MIL, CSA, etc.) | This could impose a completely new set of specifications. Is there a conflict between component specifications and those covering the finish product? |
| Wire | What AWG size or diameter, number of strands, material and insulation diameter and type? |
| Projected usage | Are all connection devices required simultaneously, or can the order be delivered in scheduled lots to avoid ''rush job'' pitfalls? Know your production scheduling. |
| Maximum voltage | Maximum DC and rms voltage can determine connector spacing and insulation types. |
| Manufacturing methods | Heavy duty or especially constructed connection devices may be required to withstand your production processes. |

The common selection factors for printed board connectors can be summarized to include:

- Contact termination, i.e., soldered through-hole or surface mount, solderless (wire) wrap, solderless press-fit, crimped or solderless IDC [9, 10]
- Number of contacts, usually a maximum of 100 for edge-board connectors, with several hundred in two-part versions
- Contact spacing, usually 2.54mm (0.100 inch) with some two-part versions having closer spacings
- Rows of contacts, one or two for edge-board and up to four for two-part connectors
- Insertion and withdrawal force, in special applications zero-insertion forces (ZIF) edge-board connectors and low-insertion force (LIF) two-part connectors are available
- Plug-in printed board thickness, usually 1.6mm (0.062 inch) for edge-board connectors
- Mating connector mounting, free-standing or backplane. [11]

### 3.2.1.1 Edge-Board (One-Part) Connectors
The edge-board type of printed board connector, Figure 3.6, consists of a number of contacts that are held in place within an insulating body in such a manner that they can mate with edge-board contacts that are an extension of the conductive pattern on the rigid printed board. The printed board can be single-sided, double-sided, or multilayer and each connector contact may engage either one side (single-readout) or both sides (double-readout) of the board. The contact has either a cantilevered spring or a bifurcated-bellows configuration.

**FIGURE 3.6.**    Edge-board (one-part) printed board connector. (*Courtesy of Continental Connector Corp.*)

The main advantages to using edge-board, as compared to using two-part connectors include reduced size and weight, reduced connector procurement costs, and the unique ability to have either single- or dual-readouts.

The disadvantages include higher printed board fabrication costs due to the need for the more rigid manufacturing controls and procedures needed to provide adequate/special plating quality (usually gold), and the mechanical registration needed for reliable (mating) operation. There is also an inherent inability of the edge-board mating connector to accommodate only a small variation in printed board thickness, bow and twist. In addition, the responsibility for satisfactory connector performance is shared by both the connector and printed board manufacturer.

### 3.2.1.2  Two-Part Connectors
The two-part type of printed board connector consists of plug and receptacle connector pairs, such as those shown in Figure 3.7. One half of the connector pair is almost always terminated to a plug-in (daughter) printed board assembly. The other half is either terminated to a fixed-position (mother) printed board assembly of several connectors, as shown in Figure 3.7, or individually hardware-mounted in a printed board (card) rack or chassis assembly.

Two-part connector contacts are made in several configurations. The most common is the box-and-post type that is used in the DIN and reverse-DIN (female daughter board/male backplane) applications. Other contact combinations are the pin-and-socket, blade-and-fork, and hermaphroditic (sexless) types.

The main advantage of using two-part, instead of edge-board, connectors is that they perform more consistently where reliability and environmental stability are required. This is because two-part connectors can be procured as self-contained mated pairs that are less affected by the assembly system to which they are attached. Thus only two-part connectors are used for military applications. The applicable specifications are MIL-C-55302 and MIL-C-83503 (insulation displacement).

Another advantage is that the board fabrication costs are usually lower and a variety of board thicknesses, such as arise with multilayer boards, can be more

**FIGURE 3.7.**    Two-part (reverse-DIN) printed board connector assemblies. (*Courtesy of Teradyne Connection Systems, Inc.*)

easily accommodated. Conversely, the main disadvantage of using two-part connector systems is the need for the use of two distinct connector parts. This results in having added connector procurement and assembly costs.

A wide selection of special two-part connectors that provide variations of the conventional printed board connectors and cabling is available for special applications. Special contact variations include the use of coaxial and fiber-optic contacts. Special connectors systems based on the use of conductive elastomers interconnection devices are gaining in popularity. [12]

### 3.2.2    Electronic Component Sockets

The use of electromechanical component sockets is an important aspect of many electronic packaging applications. With the ever increasing sophistication (and cost) of integrated circuit components, the added procurement cost of the socket can often be compensated for by reduced assembly and maintenance costs. However, serious consideration must be given to the potential reduction in performance and reliability associated with their use. [13]

#### 3.2.2.1    *Dual-Inline Package Sockets*

Dual-inline package (DIP) sockets are available in a wide range of sizes, Figure 3.8, to meet interconnection needs that range from 6 through 40 positions. DIP sockets usually have replaceable one-part sockets that give a dual-wiping action on the flat component leads.

#### 3.2.2.2    *Chip Carrier Sockets*

Sockets for use with both leaded and leadless chip carriers, Figure 3.9, are being used in both surface-mounting and through-hole assembly applications. Such devices commonly have 68 I/O's on staggered 2.54mm (0.100 inch) centers, although other terminal counts and spacings are gaining in popularity.

#### 3.2.2.3    *Pin-Grid Array Sockets*

Sockets are also available for use with pin-grid array (PGA) component packages. PGA sockets are suited for terminal counts of one hundred or more. PGA sockets commonly have leads on 2.54mm (0.100 inch) centers to offer com-

**FIGURE 3.8.**    Automatic machine insertable dual-inline package (DIP) sockets. (*Courtesy of Precicontact, Inc.*)

**FIGURE 3.9.**    Leadless chip carrier socket. (*Courtesy of Augat Inc.*)

patibility with the supplementary use of DIPs and through-hole technology. In order to overcome the high insertion forces that are normally associated with the mating of a large number of PGA pins with socket contacts, zero insertion force (ZIF) pin-grid array sockets, Figure 3.10, are often used.

### 3.2.3  Bus Bars

Since the early days of the use of integrated circuits on printed wiring boards, the bus bar has been used in many electronic circuit packaging applications. Basically, a bus bar consists of a group of conductors that carry electric current. Thus, it is generally used to meet the basic requirements of power/ground distribution systems, such as are found on printed wiring boards, backplanes, and cabinet cabling. [14, 15] For these applications, Figure 3.11, bus bars can be categorized as being either:

- Single- or multi-conductor printed wiring board fingers of strip bus bars, stiffeners, or shields
- Backplane clip-on, single- and multi-conductor bus bars

**FIGURE 3.10.**    Zero-insertion force (ZIF) pig-grid array (PGA) socket. (*Courtesy of AMP, Inc.*)

- Single- or multi-conductor laminated system bus bars
- Special bus bars, including jumpers, shorting devices and other configurations.

Originally the important parameters associated with a power bus were those directly affecting its current carrying capacity, i.e., conductivity and cross-section, although other factors such as emissivity, shape, proximity effect, and skin effect were also considered. Recently, however, other factors, including high-frequency capabilities, are becoming primary concerns. The high-frequency concerns stem largely from the high signal-transition rates that are characteristic of modern electronic equipment. This, in turn, has led to the consideration of the bus bar as a circuit component that has the attributes of a low-pass filter.

### 3.2.3.1  Printed Wiring Board Bus Bars

Printed wiring board bus bars are either of the vertical (strip) or horizontal (strip, finger, or window) type. The use of each type presents relative advantages and disadvantages for the circuit packaging engineer.

The vertical strip printed board bus bar consists of a single insulated or uninsulated layer, or multiple laminated layers of each type of layer. Its basic

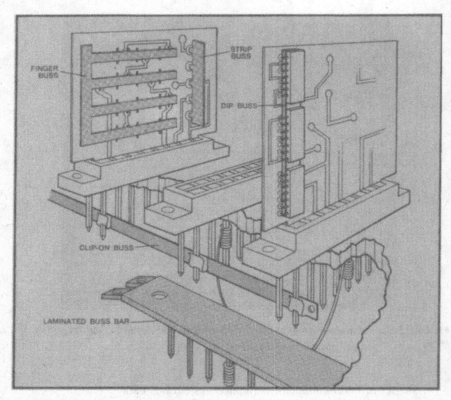

**FIGURE 3.11.**    Electronic component bus bars. (*Courtesy of Bussco Engineering, Inc.*)

function is as a power/ground distribution device that is used with the intention of eliminating localized power/ground distribution conductors on the board. When used in this manner, the strip bus minimizes the complexity of the interconnection wiring on the printed board such that a double-sided board can be used where a multilayer board would be otherwise required. It can also be used for capacitance, mechanical stiffening and shielding purposes.

The horizontal bus bar is usually associated with the use of dual-inline package (DIP) integrated circuits. The DIP bus is usually a multilayer power and ground distribution device. As with the vertical strip bus bar, the DIP bus bar can be used for a row of components or used over the entire surface of the board in a finger or comb configuration to provide multilayer interconnection capability with a double-sided board assembly. Other advantages include the ability of the bus bar to provide for local heat dissipation for the DIPs that are straddle mounted above it and to improve the electrical performance of the integrated circuits.

Window versions of the horizontal printed board bus bar distribute power and ground around all of the DIP integrated circuits in order to create a power distribution field. Because of their basic design, window bus bars can also serve as board stiffeners and electrical shields.

### 3.2.3.2  Backplane (Clip-On) Bus Bars

Formed metal clip-on bus bars are generally used to provide power distribution along rows of solderless (wire) wrap posts that extend through a printed wiring backplane or connector mounting plate. Unlike the printed wiring bus bars, the clip-on types have the advantages of being readily attached (or removed) to the connector posts in a solderless manner that minimizes contact resistance for either high- or low-current distribution purposes. However, each strip can only accommodate either power or ground interconnections, and multiple row finger or comb patterns are not feasible.

### 3.2.3.3  System Bus Bars

Laminated system bus bars are used to provide the main power distribution between several backplane assemblies and the other units in the system, including power supplies and control panels. These bus bars should be customized because of the higher currents involved and the particular physical relationships among the electronic assemblies in the equipment. By providing specifically configured layers of conductors and insulators, these bus bars can efficiently replace the use of discrete wire and cable power distribution harnesses.

## 3.3  MECHANICAL CIRCUIT COMPONENT HARDWARE

For obvious reasons, electronic circuit components and interconnection devices perform a vital function. Often the function of the supporting mechanical components (hardware) is not as obvious. For example, such components can be used to provide component-mounting devices that mechanically support critical or especially heavy circuit components such as capacitors, or they can be in the form of heat dissipaters (sinks) that play a critical role in the thermal management of the circuit performance.

### 3.3.1  Component-Mounting Devices

The environmental shock and vibration disturbances to which an electronic assembly may be subjected during normal handling, testing, and end product application can damage circuit components. When such damage is possible, the potentially affected components should be mechanically secured in place with devices such as clips/clamps, straps, or adhesives.

Metal and plastic electronic circuit clips and clamps are available to facilitates the shock and vibration protection of circuit components. They are usually available to accommodate a particular type and size of component, or they can be designed to handle a range of components.

An elastic strap can be wrapped over the body of a circuit component and through holes in the mounting base to provide a degree of mechanical retention that is relatively independent of the component's size or shape. Such straps are less expensive and easier to install than are most clips or clamps and they provide a more resilient mounting for the component. In some applications more than one strap is used, especially for larger/heavier components and for more extended environmental disturbances.

Adhesives can be used to permanently hold components in place. The permanency of the retention can be an advantage or a disadvantage depending on the application. However, unlike the use of clips, clamps or straps, care must be taken when using adhesives to be sure that they are only used in the proper amounts in the locations designated for their use and that they do not degrade the cleanliness of the circuitry on the assembly.

### 3.3.2  Heat Dissipaters (Sinks)

The increasing density and power dissipation of electronic circuit components sometimes necessitates the use of discrete metal heat sinks. Such devices can be used to provide localized heat dissipation or they can be part of an overall equipment thermal management system. (As will be discussed in Chapter 4, printed wiring boards with integral thermally conductive metal cores can also be used for this purpose.)

Heat sinks range in complexity from those that can be in the form of simple, flat metal plates to which the component is attached, discrete finned machined metal heat dissipaters (Figure 3.12), finned multiple-component extruded metal heat sinks, liquid filled vessels (blankets), to systems that use thermal feedback sensors to control liquid coolant flowing though tubes and heat pipes.

The proper selection of a heat dissipater is dependent on several considerations. These factors are interrelated and thus should be carefully considered. Of particular concern are:

- The amount of power emanating from the electronic circuit component that is to be dissipated as heat
- The maximum allowable component junction temperature and the amount of temperature rise between the junction of the component and the outside of its package (case) for a unit of power dissipation (coefficient "theta junction-to-case")
- The physical size and shape of the component and the nature of the surface

**FIGURE 3.12.**    Printed wiring board discrete-component heat sinks. (*Courtesy of Thermalloy Inc.*)

from which the heat is being dissipated, i.e., metal or plastic, rough or smooth, etc.

- The space available for the heat sink and its orientation within the equipment, i.e., vertical or horizontal
- Worst-case ambient air temperature and flow conditions, i.e. natural or forced convection
- The intended method for mounting the component and heat dissipater
- The required heat sink performance as expressed by the amount of temperature rise between the case of the component and the outside ambient environment for a unit of power dissipation (coefficient "theta case-to-ambient").

Thus, for practical purposes, the selection of a heat dissipater must be a compromise. The trade-offs being among thermal efficiency, cost, size/weight, and ease of assembly.

## References

1. Byron Johnson and Robert C. Luthi, Chinteik International (USA), Inc., "A Review of Integrated Circuit Packaging Options," *Microelectronic Manufacturing and Testing*, March 1989, pp. 12–13.
2. "Printed Board Component Mounting," IPC-CM-770, Revision C, Institute for Interconnecting and Packaging *Electronic Circuits*, January 1987.
3. Karl W. Rosengarth, Jr. and Ernel R. Winkler, Motorola, Inc. "Surface Mounting Fine-Pitch Chip Carriers," *Electronic Packaging & Production*, January, 1986, pp. 121–123.
4. Harvey R. Waltersdorf, Thomas & Betts Corp., "Choosing Packages Wisely Pays Off in I/O, Speed, Space," Reprinted with permission from *Electronic Design*, Vol. 32, No. 6, June 19, 1986, pp. 107–111.
5. Roy Richards, Fujitsu Microelectronics, Inc., "Trends in Semiconductor Packaging," *Electronic Products*, November 1989, pp. 59–62.
6. "Component Packaging and Interconnecting with Emphasis on Surface Mounting," IPC-CM-780, Institute for Interconnecting and Packaging Electronic Circuits, March 1988.
7. Larry Bos, Wink Winkelmann and Buck Robbins, IRC Inc., "An Overview of SMT Resistor Packaging," *Surface Mount Technology*, October 1989, pp. 69–71.
8. Gerald L. Ginsberg, Component Data Associates, Inc., "Connectors: EPP Tutorial Series," Electronic Packaging & Production, July 1982, pp. 233–248.
9. Gerald L. Ginsberg, Component Data Associates, Inc., "IDC — A Mature Technology Continues to Grow," *Electronic Packaging & Production*, May 1983, pp. 57–60.
10. Gerald L. Ginsberg, Component Data Associates, Inc., "Surface Mounting Impacts Printed Wiring Connectors," *Electronic Packaging & Production*, November 1984, pp. 96–99.
11. Gerald L. Ginsberg, Component Data Associates, Inc., "Printed Wiring Backplanes Reach High Performance Levels," *Electronic Packaging & Production*, April 1985, pp. 48–53.
12. Gerald L. Ginsberg, Component Data Associates, Inc., "Connectors Link Fiber Optics to a Bright Future," *Electronic Packaging & Production*, May 1983, pp. 95–97.
13. Gerald L. Ginsberg, Component Data Associates, Inc., "Connector Innovations Reflect VLSI Popularity," *Electronic Packaging & Production*, December 1983, pp. 107–108.
14. Terry Parks, Logic Dynamics Inc., "Bus Bars for PCB Applications Keyed to Design, Performance," *Electri-Onics*, June 1984, pp. 23–26.
15. Dallas Erickson, Associate Editor, "Bus Bar Applications," *Electronic Packaging & Production*, August 1981, pp. 68–91.

# 4

# Multiple Bare-Die Packaging Technologies

There are several different ways to package and interconnect integrated circuits. Traditionally, integrated-circuit dice have been used in large quantities in individual packages supplied by the device manufacturer.

In many instances the shift from through-hole to surface-mount technology has produced cost-effective packaged integrated circuit solutions for many applications. However surface mounting is only a beginning.

Other more appropriate, cost-effective, high-speed/high-density packaging technologies are required. Thus, the demand for enhanced electronic product performance has increased the use of multiple bare-die assemblies that are based on the use of hybrid microcircuit, chip-on-board (COB), microminiature multichip module (MCMs), and tape automated bonding (TAB) technologies.

## 4.1 BARE-DIE TERMINATION TECHNIQUES [1, 2]

A few different discrete integrated circuit packaging techniques have been developed for use in multiple bare-die assemblies. These have been tailored to the point where they are suitable for use in assemblies with either laminated printed board, ceramic, or silicon substrates. The basic bare-die attachment and termination techniques, or derivatives thereof, are those traditionally associated with wirebonding (Chip-and-Wire) and reflow soldering.

### 4.1.1 Chip-and-Wire Bonding

Chip-and-Wire technology is primarily of three different types, i.e, thermocompression, ultrasonic, and thermosonic. They derive their names from the

67

method the energy source employed to terminate very small (0.0025 mm (0.001 inch diameter)) gold or aluminum wire between the die and substrate. The basic parameters associated with the three basic wirebonding techniques are compared in Table 4.1.

### 4.1.1.1 Thermocompression Wire Bonding

As its name implies, thermocompression wire bonding uses a combination of heat and pressure to join two metals without melting. It is one of the most frequently used bonding processes. The basic method actually encompasses three different bonding process, i.e., ball, wedge, and stitch bonding.

Ball bonding is a technique suitable for high bonding rates. With this method, the wire to be bonded is fed through a capillary and an open flame or spark discharge melts the end of the wire to form a small ball. The bonding tool then forces the ball onto the land and the termination is made. Wire cutoff is achieved by a "flame-off" operation that also produces the ball used at the next wire termination.

Wedge bonding is also quite simple. It is used to produce two bonds between the wire and the land, thereby improving joint reliability. It is the oldest form of thermocompression bonding that is very useful with small diameter wires.

Thermocompression stitch wire bonding is a compromise between the wedge and ball methods. It uses a cut-off arrangement in place of the flame-off. This allows for the use of gold and aluminum wire and smaller bonding areas. The cut off process also forms the wire for the next bonding operation.

### 4.1.1.2 Ultrasonic Wire Bonding

Ultrasonic wire bonding is based on the use of a rapid ultrasonic-energy scrubbing or wiping motion in addition to pressure as the means of achieving the molecular bond. Because ultrasonic bonding can create bonds between dissim-

**TABLE 4.1.    Wire Bonding Technologies [4]**

| Wire Diameter (inch) | Attachment Method | Substrate Temperature (°C) | Comments |
|---|---|---|---|
| (Gold) 0.0007–0.001 | Thermocompression (hot substrate) | >300 | Limited repair ability |
| (Gold) 0.0007–0.002 | Thermosonic ball or wedge | 25 | Quality of bonds improved if the substrate is raised to 125°C. Can bond chips without raising substrate temperature. |
| (Aluminum) 0.0007–0.010 | Ultrasonic wedge | 25 | Low-temperature process, restricted due to tool size |

ilar materials, it is an extremely flexible process. Thus, both gold and aluminum metallurgies are compatible with this technique.

### 4.1.1.3  Thermosonic Wire Bonding

Thermosonic wire bonding combines the use of ultrasonic energy, heat and pressure to make the wire termination. It relies on the vibrations created by the ultrasonic action to scrub the bond area to remove any oxide layers and also to create the heat for bonding. Since this takes place at temperatures around 120° to 150° C, thermosonic wire bonding can be done with low-temperature materials, including gold and aluminum.

### 4.1.2  Controlled-Collapse Soldering

Controlled-collapse reflow soldering technology employs soldering directly between the integrated-circuit die "face" to the interconnecting substrate, Figure 4.1. Thus, it is often also referred to as "Flip-chip" or "Face-down" soldering.

Since controlled-collapse soldering does not use bonding wires or beam leads to make attachments to land patterns outside of the die's perimeter, it can be used to achieve the highest ratio of active silicon surface to multichip module substrate area. Thus, interconnection conductor lengths can be minimized and circuit performance maximized.

The advantages associated with the use of flip-chip technology and face-bonded dice as compared to wirebonding and TAB include:

- Faster throughput times
- More efficient use of MCM substrates

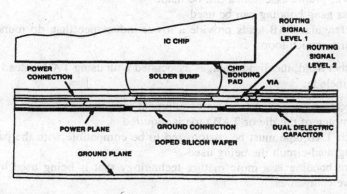

**FIGURE 4.1.**   Cross-section of flip-chip termination. (*Courtesy of AT&T Bell Laboratories.*)

- Both mechanical and electrical die attachment and final termination are accomplished in one manufacturing operation
- Circuit performance is optimized
- The number of I/Os per die is not dependent on the size of the die's perimeter.

Major disadvantages are:

- The difficulty in visually inspecting the assembled dice
- The need for precise solder bump deposition
- Difficulty in flux removal
- Potential heat transfer complications
- Limited area I/O die availability.

### 4.1.3    Tape Automated Bonding [3]

Tape automated bonding (TAB) is a "fine-pitch," bare die, electronic integrated circuit packaging technology that uses fine-line conductor patterns on a flexible printed circuit or on bare copper. The most common TAB construction is a tape carrier/interconnect product that has special features to facilitate reel-to-reel processing, Figure 4.2.

Tape automated bonding (TAB) technology is a relative newcomer to bare-die assembly applications. As compared to the use of chip-and-wire technology for the applications, TAB offers advantages that include:

- Bonding areas on the die can be hermetically sealed when bumped dice are used
- Bonding lands that can be placed closer together so that I/O counts can be much higher for the same die size
- Lower profile assemblies can be made
- Mass gang bonding can be used
- Rectangular TAB leads provide a lower inductance than do round wires with a service loop.

On the other hand, the disadvantages associated with using TAB instead of wire bonding include:

- Specially-designed fabrication equipment is needed
- Bumping of the die or TAB tape is required
- Each TAB tape must be customized to be compatible with the particular integrated-circuit die being used
- Wire bonding is a more mature technology that is being used by many more companies.

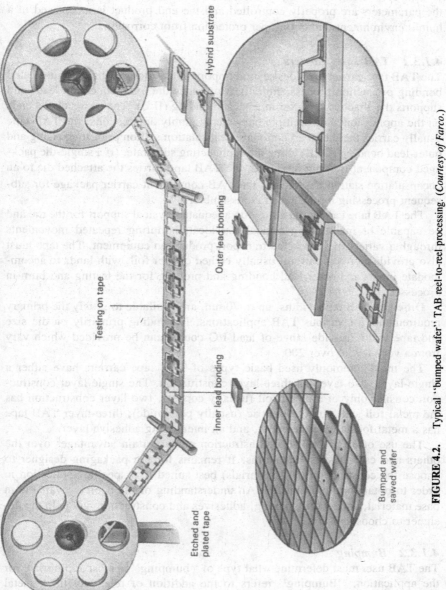

**FIGURE 4.2.** Typical "bumped wafer" TAB reel-to-reel processing. (*Courtesy of Farco.*)

Hybrid substrate

Outer lead bonding

Testing on tape

Inner lead bonding

Bumped and
sawed wafer

Etched and
plated tape

71

This combination bonding system is an excellent ILB technique when all of the parameters are properly controlled and the end product is not placed in a humid environment without proper protection from corrosion.

### 4.1.3.1  TAB Tape Carriers

The TAB tape carrier provides several important functions to the tape automated bonding production process. Initially it provides the interconnection leads and supports the attached die after inner-lead bonding (ILB). The next step depends on the application. For multiple bare-die assembly applications, the TAB tape usually carries the die to an optional encapsulation station prior to excising and outer-lead bonding (OLB) to the interconnecting substrate. (If a single-die packaged component is being produced, the TAB tape carries the attached die to an encapsulation station that creates the TAB component carrier package for subsequent processing by the substrate assembler.)

The TAB tape carrier must provide adequate physical support for the die and be capable of maintaining critical die location during repeated movements through a variety of sprocketed transport production equipment. The tape must also provide a circuit pattern, usually etched copper foil, with lands to accommodate inner- and outer-lead bonding and probing for the testing and burn-in processes.

Different TAB tape widths, up to 70mm, are available to satisfy the primary requirements for various TAB applications. Depending primarily on die size and tape width, a wide range of lead I/O counts can be provided which vary from a very few to over 200.

The most commonly used basic types of TAB tape carriers have either a single-layer, two-layer, or three-layer construction. The single-layer construction consists only of a metal foil (usually copper); two-layer construction has the metal foil and a base dielectric (usually polyimide); three-layer TAB tape has a metal foil, a base dielectric, and an intervening adhesive layer.

The use of each TAB tape construction offers certain advantages over the others and each has its limitations. It remains for the packaging designer to choose the construction (and materials) best suited to a specific application in order to obtain optimum results. An understanding of the choices available in base material, conductors, plating, adhesives and construction will help the designer to choose wisely.

### 4.1.3.2  Bumping

The TAB user must determine what type of "bumping" is most appropriate for the application. "Bumping" refers to the addition of relatively thick metal bumps to either the bonding pad sides of each individual die while they are still together on the integrated circuit fabrication wafer, or to the inner-lead bonding sites on the land pattern of the TAB tape. This is done to facilitate the inner-lead bonding (ILB) process.

Bumping of the die bonding sites on a semiconductor manufacturing wafer requires the fabrication of bonding pads (or transferred gold-plated bumps) that are raised above the planar surface of the wafer. These bumps facilitate inner-lead bonding without having the TAB tape touch the surface of the die.

A less popular alternative to wafer bumping is to form the bump on the conductive foil on the tape. This alternative is sometimes called BTAB. The most common method is to selectively etch a relatively thick copper foil on the tape to expose the bumps in the land pattern areas of the beam leads. Another approach uses semi-additive plating techniques.

### 4.1.3.3  Inner-Lead Bonding

Inner-lead bonding (ILB) is the initial assembly process for joining the inner portion of the TAB tape beam leads to the die. As applicable, ILB is followed by burn-in, testing, and encapsulation or molding of the attached semiconductor die.

The type of bond produced depends primarily on the metallurgies at the tape/die interface and the ILB method employed. Common combinations are gold-to-gold, gold-to-tin, and copper-to-copper.

Each ILB combination offers different parameters of cost, density, ease of processing, bonding force, etc., that must be matched to each particular application. For example, solder bumps offer a low cost/low bonding force option but they cannot be spaced as close together as gold bumps because of the needed volume of solder.

The termination process is based on the use of either a variation of the basic chip-and-wire termination techniques or on the use of reflow soldering principles. ILB can also have all of the leads terminated simultaneously using mass (gang) bonders or it may be done one lead at a time with "single-point" bonders.

A.  Reflow Soldering.  Reflow soldering of solder-bumped die with electroless-tin plated TAB tape have been done successfully for inner-lead bonding. In these reflow soldering applications the ILB has primarily been accomplished using a pulsed-tip thermode.

The solder used for the ILB reflow soldering operation may be any of the commonly used reflow bonding solders. However, consideration must be given to the outer-lead bonding (OLB) method when selecting the appropriate solder. Most users choose an ILB solder that has a higher melting temperature than that required for the OLB operation. A solder flux that is usually applied to the die prior to reflow is also typically used for all solder reflow operations.

Another popular ILB method combines reflow soldering and thermocompression bonding. It uses electroless tin-plated TAB tape beam leads and gold-bumped dice. The resulting bond is achieved most successfully using a heated steady state thermode.

B. Gang- and Single-Point Bonding. Mass gang bonding has been the traditional method for implementing the ILB technique with any of the processes just described. This has been primarily because of the higher throughput achieved by simultaneously bonding all of the TAB tape beam leads.

Unfortunately with mass gang bonding the flatness of the surfaces to be joined and the bonding pressures are critical. Thus, the advent of larger dice with high I/O counts has created applications that are better suited to single-point inner-lead bonding. Single-point bonding is a one-at-a-time process that uses a small wedge-like tool. It can be employed equally well for making both inner-lead and outer-lead terminations.

### 4.1.3.4 Die Protection
An important development in TAB technology involves coating or molding the die for environmental protection. This protection can be provided on automatic reel-to-reel processing machines.

General practice for bare die TAB assemblies is to apply a liquid coating to the die prior to curing the material. When a TAB component carrier package is the end product of the integrated circuit fabrication process, molding is used to encapsulate the bonded die and to provide the protective handling ring around the exposed outer TAB bonding leads.

### 4.1.3.5 Final Assembly
The final TAB assembly process begins with an excise-and-form operation that removes the die from the TAB tape. This is followed by outer-lead bonding that attaches the TAB beam leads to the interconnecting substrate.

A. Excising and Forming. Excising and forming refers to the punching out (excising) of the TAB device and the bending (forming) of its outer leads. The special, usually gullwing, shape provides a pliant lead that can withstand thermal and mechanical stresses and that still allows for a low overall assembled height.

B. Outer-Lead Bonding. The actual outer-lead bonding process is done using either a single-point or gang bonder, Figure 4.3. OLB can be done by thermocompression bonding, reflow soldering, or thermosonic bonding, depending on the metallurgical system involved. (Single-point bonding is also done with ultrasonic bonding.)

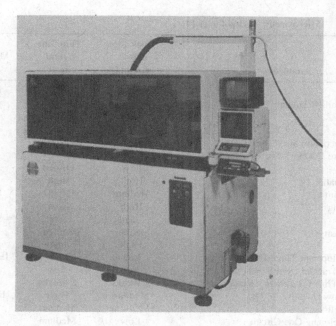

**FIGURE 4.3.**    TAB outer-lead laser bonder. (*Courtesy of Panasonic Factory Automation.*)

OLB equipment is often customized because of the wide variety of substrates involved, e.g., hybrid circuit ceramics and printed wiring boards. However, for a given class of substrates only the excising tool, thermode, lead forming tool (when required), and work fixtures need be customized.

## 4.2  HYBRID MICROCIRCUITS [4]

The use of hybrid microcircuits represents one of the approaches to bare-die electronic packaging. As the term hybrid implies, this technology encompasses multiple technical disciplines. It represents an intermediate approach between the use of conventional printed wiring assemblies with packaged components and the use of multichip modules.

The basic substrate fabrication approaches are called Thick film technology and thin film technology. Table 4.2 compares the two basic hybrid microcircuit fabrication technologies to monolithic integrated circuits.

Although the thick films are indeed usually thicker than the thin films, the terms no longer relate only to deposited film thickness. These generic terms also represent the method of film deposition, i.e., thick-film technology is based

**TABLE 4.2.  Microcircuit Technologies [4]**

| Parameter | Thick-film hybrid circuits | Thin-film hybrid circuits | Monolithic circuits |
|---|---|---|---|
| Performance | High | High | Limited |
| Design Flexibility, Digital | Medium | Medium | High |
| Analog | High | High | Low |
| Parasitics | Low | Low | High |
| Resistors, Maximum Sheet Resistivity | High | Low | High |
| Temperature Coefficient of Resistance | Low | Lowest | High |
| Tolerance | Low | Lowest | High |
| Power Dissipation | High | High | Low |
| Frequency Limit | Medium | High | Medium |
| Voltage Swing | High | High | Low |
| Size | Small | Small | Smallest |
| Package Density | Medium | Medium | High |
| Reliability | High | High | Highest |
| Circuit Development Time (Prior to Prototype) | 1 Month | 1 Month | 1–2 Months |
| 1 : 1 Design Transfer from Bench | Yes | Yes | No |
| Turnaround Time for Design Change | 2 Weeks | 2 Weeks | 1 Month |
| Part Cost, Low Quantity | Medium | High | Impactical |
| High Quantity | Medium | Medium | Low |
| Cost of Developing One Circuit | Low | Medium | High |
| Capital Outlay | Low | Medium | High |
| Production Setup and Tooling Costs | Low | Medium | High |

on the screen printing of materials upon the substrate, and thin-film technology uses vacuum deposited (sputtered or evaporated) materials.

## 4.2.1  Hybrid Substrates

Although the thick-film and thin-film hybrid microcircuit fabrication technologies are typically based on the use of different conductor and dielectric materials, in general they use similar substrate materials. Each of them can also be used with the bare-die termination technologies described in section 4.1.

The hybrid microcircuit substrate serves the same purpose as does the printed wiring board. However, unlike the printed wiring board, films of conductive, resistive, or insulative materials can be deposited directly onto the substrate to form conductors, resistors, and capacitors. To this can be added bare or packaged discrete components and integrated circuits to complete the functional electronic circuit assembly, Figure 4.4.

A variety of materials can be used for the hybrid microcircuit substrate, Table 4.3. The majority of hybrid microcircuit substrates are alumina because it represents an excellent compromise between fabrication and performance properties.

**FIGURE 4.4.**     Thin-film hybrid microcircuit. (*Courtesy of Hughes Aircraft*.)

Other ceramic materials, such as beryllia, steatite and fosterite are used to a lesser extent. In special applications, quartz and sapphire are chosen for high-frequency thin-film circuits. Where larger sizes and higher strengths are required, coated-metals are used. Other alternatives include plastics, composites, ferrites, glass and oxidized silicon.

## 4.2.2  Thick-Film Technology

Thick-film circuit elements are formed upon the substrate by the sequential screen printing, drying, and firing of the appropriate pastes. Multilayer interconnections are formed by the alternate deposition of conductive and dielectric materials.

Just as circuit requirements vary, so do the formulations of thick-film pastes vary. The typical paste contains the polycrystalline solids that are necessary to produce the desired electrical characteristics. The values of these properties control a host of parameters, such as signal propagation delay, characteristic impedance, signal risetime, and overall circuit losses. To make matters worse,

**TABLE 4.3.    Hybrid Microcircuit Substrates [4]**

| Material | Cost per 1 × 1-inch Substrate for | | | Circuit Technology Compatibility | Applications | Remarks |
|---|---|---|---|---|---|---|
| | 100 | 5000 | 100K | | | |
| Alumina | Medium | Medium | Very Low | All | Multiple | Very popular |
| Beryllia | High | Medium | Low | All | Heat Dissipating | Low thermal resistance |
| Coated Metal-Core | Medium | Low | Low | All | Special Shapes | Becomes cost effective for large substrates |
| Cofired Multilayer Ceramic | Very High | High | Medium | All | Package | May be cost effective in large quantities |
| Glass | Low | Low | Very Low | Thick and Thin Film | Limited | High thermal resistance |
| Quartz | Medium | Medium | Low | Thin Film | Saw Devices | Not popular |
| Oxidized Silicon | Medium | Low | Low | Thin Film | Multilayer Circuits | Not popular |
| Sapphire | High | High | Medium | Thin Film | VHF | Expensive |
| Ferrite | High | High | Medium | Thin Film | VHF/UHF | Very special uses |
| Organic Film | Medium | Medium | Low | Polymer Film | TAB and Multiple | Some flexible cable uses |

the material's dielectric constant and dissipation factor begin to vary significantly with increases in circuit speed and changes in temperature. Another important constituent is the glass frit that bonds the oxides and provides adherence to the substrate.

Thick-film conductors are typically silver or gold and their alloys with either platinum or palladium when the substrate is ceramic. Copper-nickel based thick-film pastes and silver filled polymers are emerging for use as low cost conductor materials.

The most widely used thick-film dielectrics are overglazes, multilayer (cross-over) dielectrics, and capacitor dielectrics; polymer dielectrics are also available for special applications. The overglaze is a low-melting-temperature vitreous glass material. A crossover dielectric is a mixture of ceramic and devitrifying glasses.

### 4.2.3  Thin-Film Technology

Thin-film circuit elements are generally formed by the photolithographic etching of resistive and conductive materials that have been deposited in a vacuum chamber on the entire surface of the substrate by either an evaporating or sputtering process.

In the evaporation deposition process the thin-film materials are heated to their vapor phase temperature and then allowed to condense onto the substrate. In the sputtering process a target is bombarded by ions that have been accelerated from a glowing discharge medium. The particles that are dislodged from the target in this manner are then deposited onto the substrate.

Gold is the primary thin-film conductor material. Copper, palladium and aluminum are also used, especially in the fabrication of monolithic integrated circuits. The primary reasons for using these alternative conductor materials are cost, conductivity and metallurgical compatibility.

The use of thin-film dielectrics are the exception, rather than the rule. When they are used they are often made from either silicon oxide, boron nitride, or silicon nitride materials. Polyimides are also used as the dielectric in very advanced applications, such as those associated with the fabrication of multichip module interconnecting substrates.

### 4.2.4  Technology Selection

Often there is only one hybrid microcircuit fabrication technology that is obviously suitable (or available) for an application. However, in an increasing number of instances a selection must be made between thick-film and thin-film technology, Table 4.4.

In general, the basic advantages of choosing the thick-film process include

**TABLE 4.4.    Cost-Effective Hybrid Microcircuit Characteristics [4]**

| Characteristic | Thick Film | Thin Film |
|---|---|---|
| Conductor Spacing | 0.010" | (0.001)° 0.0025" |
| Conductor Line Width | 0.010" | (0.001)° 0.0025" |
| Solder Pad Spacing to Component or to Other Pad | 0.020" | 0.020" |
| Resistor Tolerance | ±1% | ±0.1% |
| Resistor Temperature Coefficient | ±100 ppm/°C | ±50 ppm/°C |
| Resistor Tracking (Same Pass) | 25 ppm/°C | 1 ppm/°C |
| Resistor Ratio Match | 0.5% | 0.01% |
| Resistor Length | 0.020" | n/a |
| Resistor Width | 0.015" | (0.001)* 0.0025" |

( )*on oxidized-silicon or sapphire substrates

relatively low capital equipment requirements, a wide range in resistivities and capacitor dielectrics, and cost-effective multilayer fabrication. Limitations pertain basically to imaging characteristics (line widths and spaces) and end product circuit operating frequency.

Conversely, the relative advantages of thin-film processing include the higher degree of precision that can be achieved in the development of the circuit configuration, better component value stability, and better high-frequency characteristics. However, the disadvantage of higher fabrication equipment (and end product) costs can be an overriding consideration.

Other factors affect the comparative costs of the two technologies. Thick-film substrates do not require very critical features, whereas thin-film substrate surface finish and camber are very important. Thus, thin-film substrates are usually more expensive than the same size thick-film substrates. However, this advantage can be minimized (or negated) by the fact that the thin-film is of a higher density and, depending on component mounting considerations, a smaller substrate size might suffice.

## 4.3  CHIP-ON-BOARD TECHNOLOGY [2]

There are several ways to package and assemble integrated circuits on printed wiring boards. The newest and highest component density approach is to mount bare integrated circuit dice (chips) singly or in multiples onto a printed wiring board in what is referred to as Chip-on-Board (COB) technology, Figure 4.5.

The factors that lie behind the increasing interest in COB include its advantages with respect to:

- Significant printed wiring board land pattern size as compared to the same integrated circuit in a through-hole or surface-mount package. Also, with a lower profile, COB can be used in applications that are not possible with other traditional packaging technologies.

**FIGURE 4.5.**    Chip-on-board assembly. (*Courtesy of Valtronic Technology.*)

- Lower overall end product cost that can often be achieved by either the reduction in the size of an equivalent assembly or by reducing the number of assemblies needed to implement a circuit function. The elimination of the cost of the integrated circuit package can sometimes offset the added costs associated with COB assembly and handling.
- As with hybrid microcircuit assembly technology, advances in COB technology make it possible for any of the basic bare-die termination techniques to be used cost effectively. However, wire bonding is by far the most commonly used die-termination process, Figure 4.6.

**FIGURE 4.6.**    Ultrasonic wedge bonder. (*Courtesy of West-Bond.*)

- Improvements in the development of bare-die protection materials closely match the thermal expansion properties of printed wiring board substrates.
- The trend in semiconductor technology is for the use of highly sophisticated integrated circuits with an ever increasing number of input/output (I/O) terminations. The use of COB eliminates the concern of obtaining an appropriate package for these high I/O devices.
- The reduction in land pattern size associated with COB allows the integrated circuit dice to be mounted closer together. The resulting reduction in interconnection wiring length results in faster circuit switching speeds.
- Many, if not all, of the COB process are suitable to be automated. This can be an important consideration in what is normally a labor intensive assembly industry.

Conversely, potential disadvantages can be found with respect to:

- Fineline technology is often required for the imaging of the interconnection wiring on the printed board. Also, depending on the termination technique employed, high temperature substrate materials and gold plating might be required.
- Specialized fabrication equipment is often needed.
- The processing of assemblies with bare dice requires more care than with packaged devices.
- Die encapsulation and metallurgy must be carefully chosen to meet environmental requirements.
- Bare dice are not as readily available as are packaged devices.
- Design rules, tools and procedures may need to be modified, Table 4.5.
- COB assemblies are harder to repair than are the packaged device assembly techniques.

## 4.3.1  Chip-on-Board Substrates

The selection of an appropriate substrate material depends a great deal on the cost trade-offs associated with the end product application. For most low cost products where thermocompression wire bonding is not used, conventional printed wiring board materials are sufficient.

Where thermocompression wire bonding is employed, considerations must be given to the higher processing temperatures associated with this die termination process. For these applications high temperature epoxy/glass and polyimide/glass board materials are often used. However, when cost is an overriding consideration and cosmetic concerns can be minimized, the localized bond-site charring and delamination are often tolerated when this process is used with conventional printed wiring boards. Also, for special applications, the use of flexible printed wiring boards, conductive polymers, and molded thermoplastic board materials have been considered.

**TABLE 4.5.    Chip-on-Board Technology Guidelines [1]**

| Guidelines | Reasons |
|---|---|
| **Die-to-board attachment** | |
| Chip attachment land should be at least 0.020 in. larger than the die size on all four sides. | To allow sufficient tolerances for the chip attachment adhesive and die placement during assembly. |
| **Multichip-to-board attachment** | |
| The location of multiple chips on the board should be equally spaced and on the same axis. | To simplify die attachment, wire/lead bonding, and encapsulent/cover placement automation. |
| **Lead bonding** | |
| The bonding land on the board should be at least 0.020 in. from the chip attachment land. | To avoid bridging of the chip attachment adhesive. |
| The width of the wire/lead bonding land on the board should be at least 0.010 in. The bonding area should be at least 0.010 by 0.030 in. | To allow sufficient area for bond placment and rework. |
| **Bonding wire/lead** | |
| The length of the bonding wire/lead between the chip and board lands should not be greater than 0.100 in. | To minimize wire/lead sagging. |
| Spacing between adjacent wires/leads should be a minimum of 0.100 in. | To avoid shorts. |
| The tip of the bonding wire/lead should preferably have a square configuration. | To enable easier judgment of the reference points taken during automatic wire/lead bonding. |
| **Solder mask** | |
| The solder-mask opening should be at least 0.050 in. from the edge of the bonding lead. | To avoid bond placement on the solder mask. |

## 4.3.2   Die Attachment and Protection

As with hybrid microcircuits, depending on the die termination technique to be employed, it is often necessary to first attach the dice to the substrate.

Many COB applications require that the assembly protect the integrated circuits from the end product environment. The means for this is either the deposition of glob-top coatings or lids.

In selecting the die-protection technique to be used several important considerations must be taken into account, namely:

- The seal should be formed at a relatively low temperature and in a short enough time to minimize their effect on the die and substrate.
- The coefficient of thermal expansion (CTE) of the sealing material should be a close compromise between that of the substrate and the die in order to minimize thermal stresses and delamination.

- The seal should provide and maintain the degree of hermeticity required when exposed to the end product operating and storage environments. It should also be sufficiently stable so that it does not put excessive stress on the device termination during assembly storage and end product use.
- The sealing technique must be cost-effective, not only with respect to material costs, but also with respect to application and repairability costs.

For these reasons room-temperature vulcanized (RTV) silicone-rubber dispersion coatings are employed for COB applications. Special controlled-technology silicone gels and epoxy coatings are also used. When metal lids are chosen, they are usually soldered to the printed wiring board during the final assembly operation.

## 4.4  Multichip Modules [5]

Multichip modules (MCMs) are attractive for those system applications where interchip delays are critical to performance. The use of MCMs is also advantageous for those applications where a functional unit can be clearly defined.

In addition to conventional ceramics and glass epoxies, new substrate materials are being used, including silicon wafers. In conjunction with this, both organic and inorganic materials are being incorporated as deposited or laminated MCM dielectric layers. Also, in some applications fine line lithography is being used with thin-film metallization to fabricate MCM circuits with very small feature sizes. However, microminiature multichip modules are more expensive per unit substrate area than are conventional packaging technologies. Thus, MCM packaging technologies place great emphasis on minimizing substrate size, while extending electronic circuit assemblies to new heights of cost-effective performance that are not otherwise achievable.

By definition, all multichip modules interconnect and package bare integrated-circuit chips (dice). Thus, an initial MCM implementation decision relates to the use of an appropriate die attachment and termination technique.

### 4.4.1  Multichip Module Types [6]

The Institute for Interconnecting and Packaging Electronic Circuits (IPC) has clarified the role of MCMs in the electronic packaging hardware hierarchy. Basically, the IPC has established the position of MCMs between those for Application-Specific Integrated Circuits (ASICs) and printed board assemblies, that it refers to as being application specific electronic assemblies.

Generically the IPC considers multichip modules to be application specific electronic subassemblies with the following general characteristics:

- Usually less than ten square inches in base substrate size

- Usually provides interconnection for surface-mounted unpackaged semi-conductor dice, that are subsequently protected by an overall coating and/or enclosure
- Several MCMs are usually interconnected by a printed board assembly
- Contain more than one discrete active integrated circuit device.

The IPC further subdivides multichip modules into three types depending on their interconnecting substrate's dielectric construction, i.e., laminated dielectric, ceramic dielectric, or deposited dielectric, and other distinguishing characteristics.

### 4.4.1.1 Laminated Dielectric Multichip Modules (MCM-L)

Laminated dielectric multichip modules, type MCM-L, are represented by products that generally use what is referred to as chip-on-board (COB) technology. They are generally characterized by:

- Constructions with laminated printed circuit board substrates
- Conductors that are almost always copper
- Conductive patterns that have been imaged either by subtractive or additive deposition processes
- Vias that are initially electrolessly-deposited copper
- When required for thermal management purposes, metal constraining cores or supporting planes.

### 4.4.1.2 Ceramic Dielectric Multichip Modules (MCM-C)

Ceramic dielectric multichip modules, type MCM-C, are represented by products that generally use what is referred to as hybrid microcircuit technology. They are generally characterized by:

- Constructions with insulating materials with a dielectric constant that is usually greater than five between the signal planes, or between the signal planes and the ground planes
- Present constructions are of ceramic or glass-ceramic alternatives
- Conductors are of fireable metal materials, such as tungsten or molybdenum, and the screen-printable frit metal thick-film conductors, such as gold, silver, palladium and copper
- Conductor widths are generally greater than 0.125 mm (0.005 inches)
- Vias are formed during the conductor screen printing operation and are of the same material as the conductors
- The number of active semiconductor devices is greater than the number of passive devices.

### 4.4.1.3 Deposited Dielectric Multichip Modules (MCM-D)

Deposited dielectric multichip modules, type MCM-D, are the newest and most specialized bare die packaging products. They are generally characterized by

- Constructions with unreinforced dielectric materials adjacent to the signal planes
- Insulating materials with dielectric constants usually less than five
- The additive dielectric is usually either deposited in situ on a supporting ceramic or metal substrate, or deposited on a metal platen and then transferred to the interconnecting substrate
- Vias are usually formed either during or plated before or after the conductor deposition process
- Copper and nickel are the most common via fill materials
- Dimensional stability is determined by the properties of the underlying substrate, which is usually either ceramic, silicon, copper, or other metals or metal composites.

### 4.4.2 Multichip Module Applications

A figure of merit that can be used to differentiate between multichip modules and conventional die packaging techniques is the efficiency rate of using the bare-die interconnecting substrate. In this approach, the total area of the semiconductor die is compared to the MCM substrate area. This relationship, expressed in percent, is sometimes called the active silicon efficiency rating.

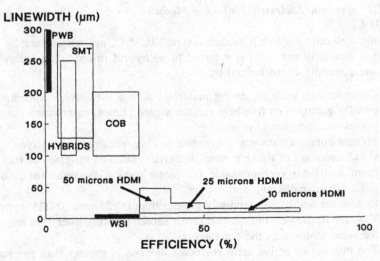

**FIGURE 4.7.**    Semiconductor/interconnecting substrate packaging efficiency. (*Courtesy of Polycon.*)

**FIGURE 4.8.**    Flip-chip termination multichip module. (Courtesy of IBM.)

The characteristic efficiency for typical die packaging technologies is shown in Figure 4.7. For example, the traditional bare-die hybrid microcircuit and chip-on-board technologies have less than a 30% active silicon efficiency, whereas all of the more advanced multichip module techniques exceed this value. Thus, examples of multichip module applications, Figures 4.8 and 4.9, are directed toward achieving a high substrate efficiency rating.

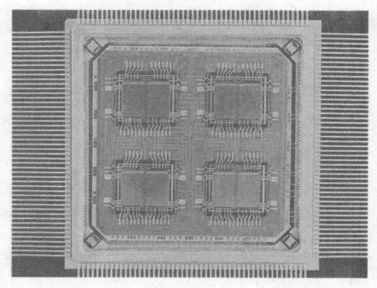

**FIGURE 4.9.**    Multiple-layer TAB multichip module. (*Courtesy of International Micro Industries.*)

**References**
1. Gerald L. Ginsberg, Component Data Associates, "Chip and Wire Technology: The Ultimate in Surface Mounting," *Electronic Packaging & Production*, August 1985, pp. 78–83.
2. Gerald L. Ginsberg, Component Data Associates, "Chip-On-Board Profits From TAB and Flip-Chip Technology," *Electronic Packaging & Production*, September 1985, pp. 140–143.
3. Gerald L. Ginsberg, Component Data Associates, "TAB Enters Multichip Modules for the Next Step in High-Density Packaging," *Electronic Packaging & Production*, October 1989, pp. 64–68.
4. Hybrid Microcircuit Design Guide, ISHM-1402/IPC-H-855, ISHM and IPC, October 1982.
5. Gerald L. Ginsberg, Component Data Associates, "Multichip Modules Gather ICs Into a Small Area," *Electronic Packaging & Production*, October 1988, pp. 48–50.
6. "An Introduction to Multichip Module Technology," Instutute for Interconnecting and Packaging Electronic Circuits, Lincolnwood, IL, IPC-TR-560, March 1990.

# 5

# Packaged Component Subassemblies

Until recently, packaged component subassembly technologies were based on the use of printed wiring boards and through-hole mounted components. However, during the last several years there has been considerable activity in developing new integrated-circuit device packages and improving the density and performance of the enhanced printed wiring boards (i.e., packaging and interconnecting structures) that mount and interconnect them.

This intense activity basically has been prompted by:

- The relatively static state of interconnection technology in the 1970s
- The emergence of new package types for the complex large-scale integration (LSI) and very-large-scale integration (VLSI) devices
- The basic packaging advantages of using surface mount attachment.

## 5.1 ELECTRICAL PERFORMANCE [1, 2]

Circuit logic switching times have traditionally decreased at a much faster rate than has the physical length of typical interconnections. This has brought about an ever increasing need for the packaging engineer to understand the importance of transmission line properties, e.g., characteristic impedance and reflections, whenever signal propagation delays due to interconnection length are significant as compared to signal rise time.

### 5.1.1 Characteristic Impedance

Characteristic impedance $Z_o$ is perhaps the key element in establishing and controlling the performance parameters of the transmission line. While the required

89

level of characteristic impedance is different for different circuit logic families, its value and control is a function of the geometry and material of the transmission environment.

Characteristic impedance is not only a function of the nominal dimensions and properties of the materials used, but it is greatly affected by the tolerances of the manufacturing processes used. Thus, for high performance systems, where close control of the signal path characteristics is critical, the interconnection packaging engineer must possess knowledge of the electronic design requirements of the end product system.

As long as the impedance of the signal path is kept relatively constant, signal reflections will not occur to a significant degree. However, where interfaces, discontinuities, or mismatches are present in the transmission path, a portion of the signal will be reflected. If this reflection is of significant amplitude, it can overcome noise immunity levels and cause a false triggering of the circuit. For this reason, an increased emphasis is being placed upon closely matching and controlling the basic elements of transmission line, including the cabling, backplanes, and connectors.

### 5.1.2  Crosstalk

Whenever a signal travels down a transmission line, its electrical and magnetic fields are concentrated in the region of the space between the signal conductor and its associated ground return path or paths. Crosstalk is the unwanted result of the linking of the fields of different signal conductors.

This crosstalk has always been a major concern at the subassembly interconnection level. Since the crosstalk is the result of capacitive and inductive coupling between the lines at high frequencies, it increases as the signal rise times decrease. It is also magnified when several parallel signal lines switch simultaneously. Thus, it is important that care is taken in the placement of the signal paths in the interconnection cabling and backplane wiring. The provision of a sufficient number of neighboring ground returns is also important.

Conventional approaches to reducing crosstalk in discrete wiring, backplanes, and connectors include:

- Minimize the length of close running parallel lines.
- Use resistive terminations on the lines of concern in order to reduce their sensitivity to signal pickup.
- Place the emitting lines as close as possible to their associated ground returns (or completely shield them) in order to reduce the magnitude of the radiated fields.
- Position the pickup lines as close as possible to their associated ground returns (or completely shield them) in order to reduce their sensitivity to stray fields by lowering its impedance.

- Move the emitting line and the pickup line as far apart as possible and, preferably, position a grounded line between them.

Transmission line concerns generally do not come into play with short lines where the propagation delay of the interconnection path is less than one-eighth of the rise time, Table 5.1. However, even for interconnections that are considered to be short, the overall inductance or capacitance may still be a matter of concern. This is because the time constant of the inductance or capacitance in the circuit environment may still be a significant fraction of the rise time.

## 5.1.3  Power Distribution

Proper power distribution is a key element in subassembly interconnection performance and reliability. However, many aspects of this factor are often overlooked by the electronic packaging engineer.

When high currents and ground are to be distributed, sometimes at multiple voltages, complex busing schemes are often required. These are often in the form of copper or aluminum bus bars (see Section 3.2.3). In these situations a few basic steps should be taken to aid in the development of an appropriate power distribution scheme, including:

- Current flows should be mapped using known power consumption data. If not, equal worst-case currents should be used.
- Power inputs and returns should be located in a manner that helps to ensure

**TABLE 5.1.    Typical Electrical Properties for Various Types of Interconnection Media**
[2]

| Interconnection type | Propagation delay | Rise length* of a 1ns rise time signal | Short-line length at 1ns rise time** |
|---|---|---|---|
| Conductors in air | 0.085 ns/in. | 11.8 in. | 1.5 in. |
| Coax cable (polyethylene dielectric) | 1.126 ns/in. | 8.0 in. | 1.0 in. |
| Ribbon cable or twisted pair (delay varies considerably) | ~0.11 ns/in. | ~9.1 in. | ~1.1 in. |
| Glass-epoxy printed board microstrip (surface) lines | ~0.14 ns/in. | ~6.9 in. | ~0.9 in. |
| stripline | ~0.18 ns/in. | ~5.4 in. | ~0.7 in. |

*Rise length is a measure of the physical spread of the rising edge of a signal travelling down a transmission line. It is defined by: rise length = rise time × signal propagation velocity.

**Short-line length is the maximum length of interconnect for which transmission line properties can be ignored. The criterion used here is: propagation delay of the line ≤ ⅛ signal rise time. Lumped inductance, capacitance, or resistance may still be of concern even for a "short" interconnection.

that current densities will be uniform and that voltage drops will be consistent with reliable operation.

- Grounding provisions should be consistent with the maximum voltage drops required.
- The interconnection wiring and interface hardware should ensure that temperature rises will be minimized.
- The interconnection wiring and interface hardware should be of a sufficient physical size and quantity to ensure that the voltage drops that do occur will not prevent the consistent delivery of the proper voltage levels at the appropriate locations in the subassembly interconnection system.

If the power distribution is less complex, basic power cabling/backplane techniques can be used. Regardless of the approach taken, the power requirements must be considered prior to the development of the overall packaging concept in order to ensure that a rational, efficient, and cost-effective system is implemented.

## 5.2  PACKAGING AND INTERCONNECTING (P&I) STRUCTURES [3]

The basic function of printed wiring boards has traditionally been to provide support for circuit components and to interconnect them electrically. To achieve the enhanced performance required by the use of packaged LSI and VLSI integrated circuits, numerous packaging and interconnecting (P&I) structure types have been developed that vary in base dielectric material, conductor type, number of conductor planes, rigidity, etc.

Since these P&I structures become the second most sophisticated component of the assembly, i.e., after the integrated circuit, the packaging engineer should be familiar with these variations and their effect on cost, component placement, wiring density, delivery cycles, and functional performance in order to select the P&I structure with the best combination of features for the particular requirements of the electronic combination apparatus or system being developed.

In general, P&I structures should be selected for optimum thermal, mechanical and electrical system reliability. However, each candidate structure has particular advantages and disadvantages when compared to the others, Table 5.2. Unfortunately, no single P&I structure will satisfy all of the needs of an application. Thus, it is important to seek a compromise of properties that is best suited for component attachment and circuit reliability.

The types of P&I structures vary from basic printed wiring boards to very sophisticated supporting-core structures. However, some selection criteria are common to all structures. To aid in the selection process, Table 5.3 lists parameters and material properties that affect system performance, regardless of P&I

**TABLE 5.2. Packaging and Interconnecting Structure Comparisons [3]**

| Type | Major advantages | Major disadvantages | Comments |
|---|---|---|---|
| Epoxy fiberglass | Substrate size, weight, reworkable, dielectric properties, conventional board processing. | Thermal conductivity, X, Y and Z axis CTE. | Because of its high X–Y plane CTE, it should be limited to environments and applications with small changes in temperature and/or small packages. |
| Polyimide fiberglass | Same as epoxy fiberglass plus high temperature X–Y axis CTE, substrate size, weight, reworkable, dielectric properties, high Tg. | Thermal conductivity, Z-axis CTE, moisture absorption. | Same as epoxy fiberglass. |
| Epoxy aramid fiber | Same as epoxy fiberglass, X–Y axis CTE, substrate size, lightest through weight, reworkable, dielectric properties. | Thermal conductivity, Z-axis CTE, resin microcracking, Z axis CTE, water absorption. | Volume fraction of fiber can be controlled to tailor X–Y CTE. Resin selection critical to reducing resin microcracks. |
| Polyimide aramid fiber | Same as epoxy aramid fiber, X-axis CTE, substrate size, weight, reworkable, dielectric properties. | Thermal conductivity, Z-axis CTE, resin microcracking, water absorption. | Same as epoxy aramid fiber. |
| Polyimide quartz (fused silica) | Same as polyimide aramid fiber, X–Y axis CTE, substrate size, weight, reworkable, dielectric properties. | Thermal conductivity, Z axis CTE, drilling, availability, cost, low resin content required. | Volume fraction of fiber can be controlled to tailor X–Y CTE. Drill wearout higher than with fiberglass. |
| Fiberglass/aramid composite fiber | Same as polyimide aramid fiber, no surface microcracks, Z axis CTE, substrate size, weight, reworkable, dielectric properties. | Thermal conductivity, X and Y axis CTE, water absorption, process solution entrapment. | Resin microcracks are confined to internal layers and cannot damage external circuitry. |
| Fiberglass/Teflon® laminates | Dielectric constant, high temperature. | Same as epoxy fiberglass, low temperature stability, thermal conductivity, X and Y axis CTE. | Suitable for high speed logic applications. Same as epoxy fiberglass. |

93

TABLE 5.2. *(Continued)*

| Type | Major advantages | Major disadvantages | Comments |
|---|---|---|---|
| Flexible dielectric | Light weight, minimal concern to CTE, configuration flexibility. | Size, cost, Z-axis expansion. | Rigid-flexible boards offer trade-off compromises. |
| Thermoplastic | 3-D configurations, low high-volume cost. | High injection-molding setup costs. | Relatively new for these applications. |
| Non-organic base Alumina (ceramic) | CTE, thermal conductivity, conventional thick-film or thin-film processing, integrated resistors. | Substrate size, rework limitations, weight, cost, brittle, dielectric constant. | Most widely used for hybrid circuit technology. |
| Supporting plane Printed board bonded to plane support (metal or non-metal) | Substrate size, reworkability, dielectric properties, conventional board processing, X–Y axis CTE, stiffness, shielding, cooling. | Weight | The thickness/CTE of the metal core can be varied along with the board thickness, to tailor the overall CTE of the composite. |
| Sequential processed board with supporting plane core | Same as board bonded to supporting plane plane. | | Same as board bonded to supporting plane. |
| Discrete wire | High-speed interconnections. Good thermal and electrical features. | Licensed process. Requires special equipment. | Same as board bonded to low-expansion metal support plane. |
| Constraining core Porcelainized copper clad invar | Same as alumina. | Reworkability, compatible thick film materials. | Thick film materials are still under development. |
| Printed board bonded with constraining metal core | Same as board bonded to low expansion metal cores, stiffness, thermal conductivity, low weight. | Cost, microcracking. | The thickness of the graphite and board can be varied to tailor the overall CTE of the composite. |
| Compliant layer structures | Substrate size, dielectric properties, X–Y axis, CTE. | Z axis CTE, thermal conductivity. | Compliant layer absorbs difference in CTE between ceramic package and substrate. |

# TABLE 5.3. Packaging and Interconnecting Structure Selection Considerations [3]

| Design parameters | Material properties | | | | | | | | |
|---|---|---|---|---|---|---|---|---|---|
| | Transition temperature | Coefficient of thermal expansion | Thermal conductivity | Tensile modulus | Flexural modulus | Dielectric constant | Volume resistivity | Surface resistivity | Moisture absorption |
| Temperature and power cycling | X | X | X | X | | | | | |
| Vibration | | | | X | X | | | | |
| Mechanical shock | | X | | X | X | | | | |
| Temperature and humidity | X | | | | | X | X | X | X |
| Power density | | | X | | | | | | |
| Chip carrier size | | X | | X | | | | | |
| Circuit density | | | | | | X | X | | |
| Circuit speed | | | | | | X | X | | |

95

structure type. Also, Table 5.4 lists the properties of the materials most common for these applications.

P&I structures can be described as being one of four basic categories of construction: organic base material, non-organic base material, supporting plane, and constraining core. However, the established reliability of the high-density, sometimes controlled-impedance, multilayer printed wiring boards made with organic-base materials has made them ideally suited for SMT usage. Thus, many of the supporting-plane and constraining-core P&I structures are enhanced versions of multilayer boards.

## 5.2.1  Organic Base Material P&I Structures

The traditional printed wiring board organic-base materials work best with through-hole mounted components and surface mounted leaded chip carriers. With leadless surface-mount chip carriers, however, the thermal expansion mismatch between package and substrate can cause problems. Flatness, rigidity, and thermal conductivity requirements may also limit their use. Finally, attention must be given to package size, input/output (I/O) count, thermal cycling stability, maximum operating temperature and solder joint compliance.

### 5.2.1.1  Epoxy-Fiberglass Base Materials
Epoxy reinforced with fiberglass is widely used in conventional P&I structures featuring through-hole, Figure 5.1, and leaded surface connections. As previously mentioned, the thermal expansion mismatch with leadless ceramic chip carriers can limit usage where the I/O count is high (above 44), where thermal or power cycling is required over large temperature extremes or a large number of cycles, or where a combination of high I/O count and thermal cycling is required.

### 5.2.1.2  Polyimide-Fiberglass Base Materials
Polyimide reinforced with fiberglass can be used where it is necessary to have improved heat resistance over epoxy-glass, and a slightly lower coefficient of thermal expansion. Depending on the environmental and functional conditions, such structures should also be limited to leaded chip carriers and leadless chip carriers of relatively low I/O count (44 or less) when thermal shock/cycling over large temperature extremes and extended life are required.

### 5.2.1.3  Epoxy Aramid Fiber Base Materials
The coefficient of thermal expansion (CTE) of epoxy reinforced with an aramid fiber, e.g., Kevlar®, closely matches that of the ceramic chip carrier, so this

**TABLE 5.4. Packaging and Interconnecting Structure Material Properties[1,2,3] [3]**

| Material | Glass transition temperature (°C) | XY coefficient of thermal expansion (PPM/°C) (note 4) | Thermal conductivity (W/M°C) | XY tensile modulus (PSI × 10^6) | Dielectric constant (At 1 MHz) | Volume resistivity (Ohms/cm) | Surface resistivity (Ohms) | Moisture absorption (Percent) |
|---|---|---|---|---|---|---|---|---|
| Epoxy fiberglass | 125 | 13–18 | 0.16 | 2.5 | 4.8 | $10^{12}$ | $10^{13}$ | 0.10 |
| Polyimide fiberglass | 250 | 12–16 | 0.35 | 2.8 | 4.8 | $10^{14}$ | $10^{13}$ | 0.35 |
| Epoxy aramid fiber | 125 | 6–8 | 0.12 | 4.4 | 3.9 | $10^{16}$ | $10^{16}$ | 0.85 |
| Polyimide aramid fiber | 250 | 3–7 | 0.15 | 4.0 | 3.6 | $10^{12}$ | $10^{12}$ | 1.50 |
| Polyimide quartz | 250 | 6–8 | 0.30 | 0.2 | 4.0 | $10^{9}$ | $10^{8}$ | 0.50 |
| Fiberglass/teflon | 75 | 20 | 0.26 | 3–4 | 2.3 | $10^{10}$ | $10^{11}$ | 1.10 |
| Thermoplastic resin | 190 | 25–30 | | | $10^{17}$ | $10^{13}$ | NA | |
| Alumina-berylia | NA | 5–7 21.0 | 44.0 | 8.0 | $10^{14}$ | | | NA |
| Aluminum (6061 T-6) | NA | 23.6 | 200 | 10 | NA | $10^{6}$ | | |
| Copper (CDA 101) | NA | 17.3 | 400 | 17 | NA | $10^{6}$ | | |
| Copper-clad invar | NA | 3–6 | 150XY/20Z | 17–22 | NA | $10^{6}$ | — | NA |
| Copper-clad molybdenum 13%/74%/13% | NA | 5.7 | 209XY/161Z | 43.5 | NA | $10^{6}$ | — | NA |
| Copper-clad molybdenum 20%/60%/20% | NA | 6.9 | 244XY/195Z | 41 | NA | $10^{6}$ | | NA |
| Graphite (P-100) | 160 | −1.15 | 110XY/1.1Z | 37 | NA | | | |
| Graphite (P-75) | 160 | −0.97 | 40XY/.7Z | 25 | NA | | | |

Notes: 1. The materials can be tailored to provide a wide variety of material properties based on resins, core materials, core thickness, and processing methods.
2. The X and Y expansion is controlled by the core material and only the Z axis is free to expand unrestrained. Where the Tg will be the same as the reinforced resin system used.
3. When used, a compliant layer will conform to the CTE of the base material and to the ceramic component, therefore reducing the strain between the component and pkd structure.
4. Figures are below glass transition temperature, are dependent on method of measurement and percentage of resin content.
NA = Not Applicable.

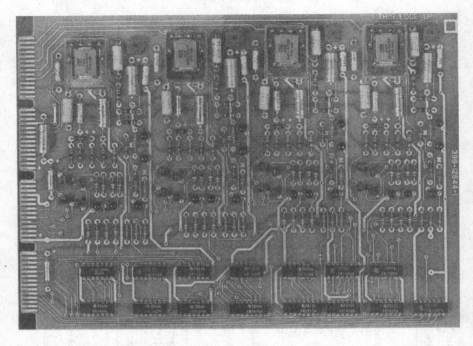

**FIGURE 5.1.**    Organic-substrate through-hole component mounting assembly. (*Courtesy of Philco-Ford Corp.*)

type of structure is appropriate for use with high-I/O count chip carrier packages and temperature cycling over wide temperature extremes. Because of stresses that are generated between the fibers and resin matrix during temperature cycling, microcracks in the surface of the substrate have been found. These microcracks are minimized with certain resin systems.

Water absorption is higher than with epoxy fiberglass materials, and could cause processing problems if not removed before use. Water absorption in field environments could also be a problem unless some sealing or coating method is used.

### 5.2.1.4  Polyimide Aramid Fiber Base Materials
Higher heat resistance makes the use of aramid-fiber reinforced polyimide laminates slightly better for rework with heated tools than the epoxy materials, as well as permitting temperature cycling over greater extremes. Its low Z-axis expansion also improves the reliability of through vias in the substrate. Because of its low CTE, it is ideally suited for high I/O count packages and where extreme environments and long life are prerequisites. However, these materials have problems with water absorption and microcracking.

### 5.2.1.5  Polyimide Quartz Materials

Fused silica (quartz) reinforced polyimide resin also has a CTE that closely matches that of the ceramic component. By varying the volume fractions of fabric and resin, the CTE can be varied. However, the amount of resin must remain below 40% to achieve a CTE of 6-8 PPM/degrees C. Slight changes in resin content and/or layers of copper drastically affect the overall CTE of the P&I structure.

Laminates reinforced with quartz are extremely abrasive, and thus shorten drill life. New drill design and development may ease this problem. Fabric availability is also a problem.

### 5.2.1.6  Fiberglass/Aramid Fiber Materials

Fiberglass reinforced resin systems can be combined with an aramid fiber reinforced resin system to produce a low expansion composite material. The volume fraction of the aramid fiber can be adjusted to match the CTE value of the ceramic component. The aramid fiber reinforced material may be either on internal or external layers.

When the aramid fiber material is confined to the internal layers, the resin microcracking is also contained. This eliminates surface cracks that could create reliability problems such as cracked conductors, while improving the machinability and drilling of the aramid fiber containing the P&I material. Its tailorable CTE makes it suited for all leadless applications.

### 5.2.1.7  Teflon® Fiberglass Materials

Teflon reinforced with fiberglass is used almost exclusively in radio-frequency (RF) applications because of its low dielectric constant. Also, it is only recommended for use with small leadless components because of its high expansion coefficient. Single-sided and doubled-sided constructions are common, whereas multilayer construction is rare.

### 5.2.1.8  Flexible-Dielectric Structures [4]

The use of flexible-dielectric printed circuits should be considered if it is necessary to:

- Reduce package size and weight
- Reduce assembly costs and wiring errors
- Achieve three-dimensional assembly packaging
- Improve cost effectiveness
- Enhanced flexural endurance
- Tightly control electrical performance
- Improve reliability
- Achieve compliant mounting

For these requirements there are several types of flexible printed circuits. The simplest do not require electrical interconnections between layers; the more complex do. Plated-through holes are most often used to provide interconnections between layers. The plated-through hole interconnection approach is the same as is used in conventional rigid multilayer circuits. The only difference is that in flexible circuitry the through-hole should be located in an area that will receive limited flexing. Accordingly, it is often desirable to make rigid the areas of the circuits that contain through-holes.

A flexible printed circuit that is considered to be single-sided can take many forms beyond that which is normally considered in rigid printing wiring. For purposes of definition all flexible printed circuits that contain a single conductor layer will be considered to be of this type. Single-sided flexible printed circuits can further be divided into subclasses, such as single-access uncovered, single-access covered, double-access uncovered, and double-access covered.

A.  Single-Access Uncovered.   This type of flexible printed circuit is the most inexpensive type because it only consists of a dielectric layer that supports the conductor. This type of flexible printed circuit finds most application where the wiring will not be exposed to mechanical abuse or environmental contamination. The components are normally all mounted from the dielectric side with all electrical connections to the conductor side.

B.  Single-Access Covered.   This type of flexible printed circuit consists of a three layer structure in the form of two dielectric layers with a conductor layer sandwiched in between. The dielectric is selectively removed from those areas of the conductor to which subsequent electrical connection is to be made.

The cover lay added to the basic wiring gives the advantage of a moisture and insulation barrier for conductor to conductor spacing, and insulation from associated mounting chassis or hardware. The cover lay having oriented holes does have a significant effect on cost.

C.  Double-Access Uncovered.   This type of flexible printed circuit has the unique feature of electrical access to the single layer conductor from both sides. The structure consists of two layers, dielectric and conductor. The dielectric layer is fabricated with holes while the conductor pattern is suspended across the holes allowing contact from either side of flexible printed wiring.

Double-access uncovered flexible printed circuits have been most frequently used in high production quantity applications where inexpensive electrical connection is required with a minimum amount of components but with mounting necessary from both sides. Dedicated production tooling is usually required for this type.

D.   Double-Access Covered.   This type of flexible printing circuit is made up of three layers, i.e., two dielectric layers and a single conductor layer. The conductor layer is sandwiched between the two insulating layers which have selective access holes to the conductor. The access holes may be at opposing points, leaving the bare conductor exposed on both sides or the access areas may be selectively located on either side.

The application for double-access flexible printed circuits is the most versatile single sided type. The cover lay not only can serve the purpose of insulation from associated hardware, but can serve to insulate itself. Many applications of point to point wiring as well as component mounting have been developed with this configuration.

E.   Double-Sided Flexible Printed Circuits.   As the name implies, double-sided flexible printed circuits have conductors on both sides of a base dielectric material. Conductors on opposite sides may be connected by plated-through holes or other means. In this and other respects they are very similar in construction to rigid double-sided printed circuit boards.

Double-sided (and multilayer) flexible printed circuits should always be given serious consideration when an application's interconnections become very dense. However, in designing flexible printed circuits with more than one conductor layer, the required degree of flexibility must be kept in mind. Plating, conductor routing, and material thickness can also affect printed circuit flexibility.

F.   Multilayer Flexible Printed Circuits.   The flexible multilayer type of printed circuit is an end product consisting of three or more conductive layers on flexible dielectric bases bonded to form a monolithic mass. An insulating layer may be applied over the outside conductor paths.

The multilayer structure, in its simplest form, is a three conductor layer structure with the center layer being discreet wiring paths and both outer conductor layers being solid copper conductors. In this type of multilayer structure the wiring paths are electrically equivalent to coaxial or shielded wire configurations.

G.   Rigid-Flex Printed Circuits.   Rigid-flex printed circuits are similar to conventional multilayer printed circuit types except that the bonding and connections between layers is confined to specific areas of the interconnection wiring and component mounting plane. Between the rigid laminated areas, each conductor layer is usually bonded to a single, thin base laminate area that remains flexible even though there may be more than one conductor layer.

Thus, a rigid-flex printed circuit is generally a combination of flexible and rigid printed circuit elements that are combined through lamination and plating into a single inseparable structure. One of the most important benefits of this type of construction is a reduction in the number of discrete solder joints or connector contacts in the conductive path. Other benefits are savings in space and weight, and lower packaging assembly costs, as termination points are pre-established and eliminate the chances of interconnection wiring errors. A typical rigid-flex circuit and its application are shown in Figure 5.2.

### 5.2.1.9  Thermoplastic Resin Base Materials

An alternative to conventional organic-base printed wiring board materials for general-purpose applications is the use of high-performance engineering-grade thermoplastic resins. Typical thermoplastics for surface mount use are poly-ethersulfone (PES), polysulfone (PSF) and polyetherimide (PE).

This type of printed wiring board is well suited for high-volume applications requiring a three-dimensional P&I structure with fully-additive or semi-additive printed wiring. Copper-clad thermoplastic base materials are also available for planar subtractive printed wiring applications.

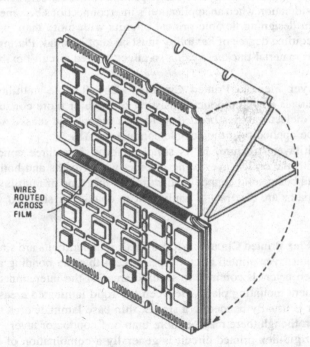

WIRES
ROUTED
ACROSS
FILM

**FIGURE 5.2.**    Rigid-flex printed board assembly. (*Courtesy of Kollmorgen Corp.*)

### 5.2.1.10  Paper-Phenolic Base Materials

Because of their relatively low, paper-phenolic and composite laminates are used in many high-volume, low-cost, packaging and interconnecting structure applications. The mechanical and electrical characteristics of these printed board materials are not as good as those of the epoxy-fiberglass substrates. However, they are available in both flame-retardent and non-flame-retardent grades. Thus, these laminates are used extensively in radio, television, and other commerical products where their coefficient of thermal expansion properties are not a major consideration.

## 5.2.2  Non-Organic Base Material P&I Structures

The ceramic (non-organic) base materials typically used with thick- or thin-film hybrid microcircuit technology (see Section 4.2) are also ideally suited for surface mount applications. They can incorporate thick- or thin-film resistors directly on the P&I structure and buried capacitor layers that increase density and improve reliability. However, repairability of the P&I structure is limited.

Ceramic materials, usually alumina, appear to be ideal for P&I structure with leadless ceramic chip carriers because of their relatively high thermal conductivity and the coefficient of thermal expansion (CTE) match. Unfortunately, the P&I structure is usually limited to a size of approximately 22,600 sq.mm (35 square inches). However, the evolving use of these materials with non-noble metals, such as copper, has attracted both military and commercial interest.

Ceramic printed circuit boards are assembled with directly soldered surface mount components and used as independent P&I structures exclusive of a hermetic package. They are usually based on the use of ceramic materials with interconnect circuit patterns which are formed using multilayer thick film/co-fired processing techniques. Ceramic boards have established reliability for high density ceramic leadless chip carrier packaging.

## 5.2.3  Supporting-Plane P&I Structures

Conventional printed wiring boards can be used with supporting metallic or nonmetallic planes or with custom processing to enhance the structure's properties. Depending on the results desired, the supporting plane can be electrically-functional or not, and can also serve as a structure stiffener, heatsink and/or CTE constraint.

### 5.2.3.1  Printed Wiring Boards Bonded to Support
### Plane (Metal or Non-Metal)
A P&I structure with controlled thermal expansion in the X and Y axes, improved rigidity, improved thermal conductivity, etc., can be formed with a con-

ventional thin printed wiring board that has been fabricated and bonded with a rigid adhesive/insulation to a supporting plane, such as metal or graphite-fiber resin composite. However, the printed wiring board must be thin enough to preclude warping of the assembly or else the board should be bonded to both sides of the plane.

The printed wiring board portion of this type of P&I structure can be either unpopulated or completely assembled and tested prior to being bonded. However, components can only be mounted to one side of the printed board. Also, the support is not normally electrically connected to the printed wiring board.

### 5.2.3.2  Sequentially-Processed Structures with Metal Support Plane

High-density, sequentially-processed, multilayer P&I structures are available with organic dielectrics of specific thickness, ultrafine conductors, and solid plated vias for layer-to-layer interconnections with thermal lands for heat transfer, all connected to a low-CTE metal support heatsink. Thus, this technology combines laminating materials, chemical processing, photolithography, metallurgy, and unique thermal transfer innovations, such that it is also appropriate for mounting and interconnecting bare integrated circuit chips.

### 5.2.3.3  Discrete-Wire Structures with Metal Support Planes

Discrete-wire P&I structures have been developed specifically for use with surface mounted components. These structures are usually built with a low-expansion metal support plane that also offers good heat dissipation.

The interconnections are made by discrete 0.06 mm (0.0025 inch) diameter insulated copper wires precisely placed on a 0.03 mm (0.0013 inch) grid by numerically-controlled machines. This geometry results in a low-profile interconnection pattern with excellent high-speed electrical characteristics and a density normally associated with thick-film technology.

The wiring is encapsulated in a compliant resin to absorb local stresses and dampen vibration. Electrical access to the conductors is by 0.25 mm (0.010 inch) diameter copper vias. The small via size can be accommodated in the component attachment land, thus eliminating the need for fan-out patterns when using components with terminals on centers as close as 0.6 mm (0.025 inch), and allowing very high packaging densities.

### 5.2.3.4  Flexible Printed Wiring Boards with Metal Support Planes

Conventional fine-line polyimide flexible printed wiring provides another arrangement for a P&I structure when it is attached to a metal support plane.

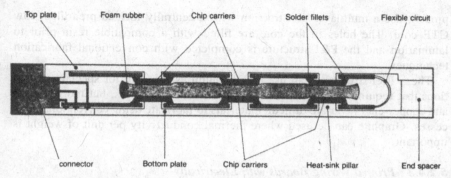

**FIGURE 5.3.**   Flexible printed wiring board assembly with external supporting planes.

These assemblies can be constructed in multilayer form while retaining the low-modulus feature that reduces residual strain at the solder joints. Furthermore, lasers can drill very fine holes in the thin, printed wiring board laminate. These holes can be plated-through or filled with solid copper, as required.

Cutouts can be made in the flexible circuit to accommodate pillars from the metal heatsink support plane in order to retain inherent flexibility while dissipating heat from the solder joint, Figure 5.3. Although this appears to be heavy and cumbersome, if the heatsink base-plates are made from thin sheets of aluminum, the resulting density of the combined circuit/heatsink assembly might actually be less than other constructions.

### 5.2.4   Constraining-Core P&I Structures

As with supporting-plane P&I structures, one or more supporting metallic or nonmetallic planes can serve as a stiffener, heatsink, and/or CTE constraint in constraining-core P&I structures.

#### 5.2.4.1   Porcelainized-Metal (Metal Core) Structures

An integral core of low-expansion metal (e.g., copper-clad Invar), can reduce the CTE of porcelainized-metal P&I structures so that it closely matches the CTE of the ceramic chip carrier. Also, the P&I structure size is virtually unlimited. However, the low melting point of the porcelain requires a low-firing-temperature conductor, and dielectric and resistor inks.

#### 5.2.4.2   Printed Wiring Boards with Constraining (not Electrically-Functional) Cores

Printed wiring boards bonded back-to-back to a constraining core can be used for high-density, low-warpage P&I structures. The core acts as a heatsink, but in this case is not electrically functional. For optimum density with this approach, use a multilayer construction with a centrally located predrilled, low CTE core. The holes in the core are filled with a compatible resin prior to lamination and the P&I structure is completed with conventional fabrication techniques.

These P&I structures can use molybdenum as the core for special applications that require inherent stiffness in extreme environments, but molybdenum and copper-clad Invar are difficult materials to fabricate using conventional processes. Graphite can be used where thermal conductivity per unit of weight is important.

### 5.2.4.3  Printed Wiring Boards with Electrically-Functional Constraining Cores

Enhanced performance multilayer printed wiring boards can be made as P&I structures with thin, 0.1 to 0.25 mm (0.004 to 0.010 inch), copper-clad Invar being used as electrically-functional ground and power planes. The constraining cores can be in the form of clad laminate or predrilled uninsulated planes. Such planes should be located in a symmetrical arrangement within the multilayer lay-up and subsequently laminated as an integral part of the P&I structure. The overall CTE of the structure can be tailored by varying the composition and thickness of the planes.

### 5.2.4.4  Printed Wiring Boards with Constraining Cores

X- and Y-axis thermal expansion, rigidity, and thermal conductivity can be improved with the use of a constraining fiber resin composite internal plane in a conventional printed wiring board which can be modified depending on the properties of the supporting plane and its location within the P&I structure. These constraining fibers can be graphite, aramid fiber, quartz, etc. The very high modulus of these materials requires a balanced construction to prevent bowing or twisting.

### 5.2.4.5  Compliant-Layer Structures

Compliant layer P&I structures were developed specifically for use with chip carriers and surface mounting. A novel elastomeric first layer provides cushioning and minimizes the risk of solder-joint failure due to differential thermal expansion between the structure and large surface-mounted chip carriers.

## 5.3  GENERAL ASSEMBLY SELECTION CONSIDERATIONS

The selection of an appropriate packaged-component assembly technique should initially include the requirements of the end product equipment and subassembly from the viewpoint of form, fit and function with respect to cost effective-

**TABLE 5.5.    Integrated Circuit Packaging Technology Comparison [5]**

| Characteristics | Through-hole | Leaded surface mount | Leadless surface mount | Bare die |
|---|---|---|---|---|
| Packaging density | Low | Moderate | Good | High |
| Standardization | Very good | Good | Good | Limited |
| Thermal performance | Moderate | Good | Very good | Fair |
| Substrate choices | Very good | Very good | Good | Limited |
| Fab investment | Low | Low | Moderate | High |
| Assembly investment | Moderate | Low | Moderate | High |
| Support investment | Low | Moderate | Moderate | High |
| External assembly services | Very high | High | Moderate | Limited |
| Maintenance skills | Low· | Moderate | Moderate | High |
| Change risk | Very low | Low | Moderate | High |
| Assembly test | Very easy | Easy | Moderate | Complex |
| Documentation | Easy | Easy | Moderate | Complex |
| Logistics support | Field change | Field change | Field change | Factory only |
| Inspect (circuit) | 100% test | 100% test | 100% test | Lot accept |
| Component burn-in | Easy | Easy | Easy | Impractical |
| Pretest | Very easy | Very easy | Easy | Impractical |
| Change and repair | Easy | Easy | Easy | Difficult |
| Component availability | Excellent | Very good | Good | Limited |
| Multiple sourcing | Excellent | Very good | Good | Limited |
| Footprint commonality | Excellent | Very good | Good | Poor |
| Profile | High | High | Moderate | Low |

ness, performance, and marketability issues. After characterizing the assembly in this manner, it is then necessary to select an implementation technique based on specific electrical and mechanical functions.

Determining factors will include packaging density, assembly profile height, development time, development cost, circuit element factors, manufacturing costs, thermal considerations, reliability, etc., and specific related implementation details, Table 5.5.

The assembly process steps differ according to the type of product being assembled, i.e., through-hole, surface mount, or mixed technology. They also vary according to manufacturer expertise, experience and preference. Table 5.6 compares some of the possible process flow sequences for various types of products.

## 5.4  THROUGH-HOLE TECHNOLOGY (TMT) [5]

Through-hole assembly technology is based on the use of packaged components such as dual-inline packages (DIPs), single-inline packages (SIPs), and other

**TABLE 5.6. Integrated Circuit Assembly Process Flow Comparison [6]**

| Through board | Surface mount single sided | Surface mount single sided | Surface mount double sided | Surface mount through board mix | Chip on board TAB | Chip on board |
|---|---|---|---|---|---|---|
| Preclean | Preclean | Preclean | Preclean | Preclean | Preclean | Preclean |
| Auto insert components | Apply solder cream | Apply comp. adhesive | Apply solder cream side 1 | Apply solder cream | Apply die att. adhesive | Apply die att. adhesive |
| Manual insert components | Auto pick/place components | Auto pick/place components | Apply adhesive side 1 | Auto pick/place components | Auto place tab chip | Auto place chip |
|  | Manual pick/place components |  | Auto pick/place components side 1 | Manual pick/place components |  |  |
|  | Cure solder cream |  | Cure solder cream | Cure solder cream | Cure adhesive | Cure adhesive |
|  | Flow melt solder cream |  | Flow melt solder cream | Flow melt solder cream |  |  |
|  |  |  | Clean | Clean |  |  |

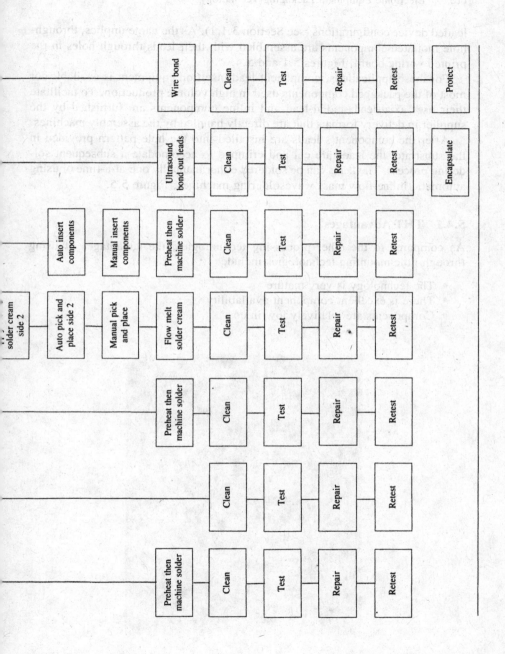

leaded device configurations (see Section 3.1.1). As the name implies, through-hole mounted components are assembled with their leads through holes in the printed wiring board, Figures 5.1 and 5.4.

For these applications, component-lead insertion equipment is available for most of the packaged components used in high volume production. To facilitate their use, axial-lead, radial-lead and inline components are furnished by the supplier in delivery formats that are directly handled by the assembly machines.

After the component's leads are installed into the hole pattern provided in the substrate, the leads are cut and crimped to accomodate a subsequent soldering process. The leads can be soldered either manually one-at-a-time or using automatic inline-flow mass wavesoldering machines, Figure 5.5.

### 5.4.1  THT Advantages

As compared to the other processing technologies, the advantages of using through-hole mounting technologies include:

- The technology is very mature
- There is excellent component availability
- Components are relatively low in cost

**FIGURE 5.4.**    Multilayer through-hole printed wiring board assembly. (*Courtesy of Ford Aerospace Corp.*)

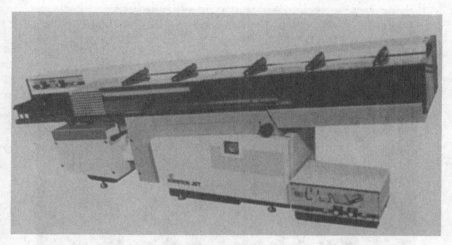

**FIGURE 5.5.**    Jet wave soldering machine. (*Courtesy of Kirsten Kabeltechnik AG.*)

- Computer-aided tools are available to support design, fabrication, assembly and testing operations
- End-product reliability is well established
- Electrical test needs are well understood
- There are existing assembly test procedures
- Rework capabilities exist.

### 5.4.2  THT Disadvantages

Conversely, disadvantages pertain specifically to:

- Component mounting density is minimal
- End-product weight is increased
- Component lead count is limited (except with pin-grid array packages)
- Component mounting holes minimize interconnecting conductor routing density

## 5.5  SURFACE MOUNT TECHNOLOGY (SMT) [3]

The development of through-hole packaged-component assembly technology has essentially reached its limits as far as improvements in end-product cost, packaging density, circuit performance and reliability are concerned. To realize further benefits in these areas the trend is toward the increased use of surface mount technology (SMT) or a mixed SMT/through-hole technology. As the

**FIGURE 5.6.**    Multilayer surface mount printed wiring board assembly. (*Courtesy of Boeing Aerospace Corp.*)

name implies SMT terminates packaged circuit components in a planar manner on the surface of the printed wiring board, Figure 5.6.

### 5.5.1  SMT Advantages

The increased usage of SMT has had a profound impact on packaging technology because of its use of relatively smaller component packages and the fact that it does not require component mounting holes in the interconnecting substrates. This translates into packaging advantages such as:

- Reduced subassembly, and ultimately equipment, size and volume through the use of increased component placement densities, finer-pitch component terminals, and the ability to have components mounted on both sides of the assembly.
- Reduced component mounting costs by eliminating the need for packaged-component lead forming. This also facilitates the increase in component placement rates.
- Improved high-speed/high-frequency circuit performance with shorter interconnection wiring lengths and lower inductances, capacitances and resistances.

- Improved shock and vibration environmental stability through the use of smaller (lower mass) component packages.
- The availability of some high-speed integrated circuit devices only in surface mount configurations.

## 5.5.2  SMT Disadvantages

However this technology also has some disadvantages, including:

- As with all new packaging technologies, time is required to develop and gain experience with the necessary SMT design, manufacturing, assembly, and testing procedures.
- Some packaged circuit component types are not as readily available as are their through-hole mounted counterparts. (This will be overcome as the technology continues to mature. Also, as noted above some circuit components are only available in SMT configurations.)
- Reliability is not as readily established as it is with through-hole technology due to the relative immaturity of the technology.
- The use of adhesives is required for the attachment of surface mount components that are to be wave soldered in mixed-technology assemblies.
- Wave soldering is not suitable for use with many surface mount assembly configurations. Thus, the use of less common reflow soldering techniques, and the associated solder deposition process, is required.
- Most conventional through-hole solder joint quality criteria and inspection procedures are not applicable to SMT due to higher density and the changed nature of the solder joints.

## 5.5.3  SMT Assembly Processing [6]

As shown in Table 5.6, there are several different ways to process surface mount assemblies. Basically, the differences relate to the use of SMT components on one (single-sided) or both (double-sided) sides of the assembly, whether there is a mixed SMT/through-hole configuration, and whether or not special components must be mounted and/or soldered manually.

### 5.5.3.1  Substrate Preparation

Unlike through-hole technology, surface mounting requires additional printed wiring board substrate preparation prior to component mounting. When mixed-technology wavesoldering is used, this includes the deposition of an adhesive in the location where the wavesoldered surface components are to be placed.

One of the most significant departures from through-hole technology is the need for a solder-deposition process for all single-sided and double-sided SMT assemblies, and on the side opposite the wave of mixed technology assemblies.

Thus, the use of a solder paste (cream) is an important part of surface mount substrate preparation prior to reflow soldering. The solder paste acts partially as an adhesive before reflow and its surface tension helps to align skewed parts during soldering. It contains the flux, solvent, suspending agent, and solder that is traditionally supplied by the wave soldering machine. Therefore, the selection of a particular solder paste involves optimizing its rheological characteristics such as viscosity, flow and spread. Susceptibility to solder ball formation and wetting characteristics must also be considered.

The solder paste is generally applied on the lands of the substrate by either screening, stenciling, or syringe dispensing. The use of stencils is preferred for high-volume applications because they are more durable, easier to align, and can be used to apply a thicker layer of solder than can the screening process. However, because they are usually more expensive than screens, the use of stencils may not be suitable for low-volume production runs.

Solder preforms (doughnuts) are sometimes used for the through-hole components in mixed-technology applications in order to be able to use reflow soldering instead of wave soldering for these applications. The use of preforms and reflow soldering is particularly suitable for assemblies that consist predominantly of surface mount components.

### 5.5.3.2 SMT Component Placement
The component placement accuracy requirements for surface mounting, especially with fine-pitch component packages, often necessitates the use of automated assembly machines, Figure 5.7, that are often referred to as pick-and-place equipment. Such machines are available for either inline, simultaneous, sequential or simultaneous/sequential component placement operations.

The selection of these machines is generally based on the rate at which components are most cost-effectively assembled and their suitability for use with the appropriate variety of surface mount component package delivery formats, such as tape, stick, belt, matrix tray, and cassette feeders.

Inline pick-and-place machines employ a series of component placement stations. Each station places its respective component as the substrate moves through it down the line. Inline component placement times can vary from 1.8 to 4.5 seconds per assembly.

Simultaneous placement equipment mounts an entire array of components on the substrate at one time. Typical simultaneous component placement times vary from 7 to 10 seconds per assembly.

Sequential SMT component mounting units typically utilize a software-controlled $X$-$Y$-axis moving table system. The components are individually placed on the substrate in this type of assembly. Typical sequential component placement times vary from 0.3 to 1.8 seconds per component.

**FIGURE 5.7.**    Surface mount 'pick-and-place' assembly machine. (*Courtesy of Emhart/ Dynapert Corp.*)

Sequential/simultaneous pick-and-place machines also feature a software-controlled $X$-$Y$-axis moving table system. However, the components are individually placed on the substrate in succession from multiple component feeding heads. Simultaneous/sequential component placement times are typically about 0.2 seconds per component.

### 5.5.3.3  Soldering

The selection of a soldering process depends upon the type of components being assembled and, as previously mentioned, whether or not a mixed technology or all surface mount assembly is being manufactured. Depending on the component mix, the candidate technologies are wave soldering and the variety of reflow soldering techniques that are based on the use of either vapor-phase (condensation) energy, infrared (IR) energy, lasers, hot-belt conduction, or hot gases. Each process has a corresponding set of manufacturing parameters for which it is best suited. Another consideration is for the use of a low-volume batch processing unit or a high-volume inline machine, Figure 5.8.

**FIGURE 5.8.**    Inline vapor-phase soldering machine. (*Courtesy of Centech Corp.*)

### 5.5.3.4 Cleaning

The cleaning of surface mount assemblies is harder to perform than is cleaning with through-hole technology because of the higher density nature of the assembly, especially with respect to removing flux that has been entrapped under the component packages. This cleaning has become further complicated by the con-

**FIGURE 5.9.**    Hot-gas reflow soldering. (*Courtesy of Nu-Concept Systems Inc.*)

cerns about worldwide ozone depletion. Such concerns are eliminating the use of the traditional chlorofluorocarbon cleaning solvents.

Flux entrapment may cause potential reliability problems if the assembly is not properly cleaned. However, new solder paste formulations are being developed that might result in the use of no-clean or no flux soldering technologies in order to avoid these problems.

### 5.5.3.5 Repairing/Reworking

The repair or rework of surface mount assemblies is generally easier than with all through-hole assemblies due to the absence (or minimization with mixed technology) of component mounting holes. However, again, because of the high-density nature of the assembly, special tools are used to carefully direct the reflow soldering energy so as to minimize the amount of heat applied to the assembly during this operation. The manual component removal/replacement tools include the use of fork-like soldering iron tips and resistance-heated tweezers. Various types of hot-air devices, Figure 5.9 are also available for this purpose.

### References

1. Christopher Van Veen, Teradyne Connection Systems, "Design Considerations for Backplane Interconnection Systems," Electri-Onics, February 1986, pp. 77–80.
2. C. Michael Hayward, Hybricon Corp., "Backplane Design Considerations," Connection Technology, October 1986, pp. 35–40.
3. "Component Packaging and Interconnecting with Emphasis on Surface Mounting," ANSI/IPC-780, July 1988, Institute for Interconnecting and Packaging Electronic Circuits, Lincolnwood, IL 60646.
4. "Flexible Printed Wiring," Printed Wiring Design Guide, IPC-D-330, Section 6, Institute for Interconnecting and Packaging Electronic Circuits (IPC), Lincolnwood, IL 60646.
5. "Printed Board Component Mounting," ANSI/IPC-CM-770, Revision C, February 1987, Institute for Interconnecting and Packaging Electronic Circuits, Lincolnwood, IL 60646.
6. "Surface Mount Land Patterns (Configurations and Design Rules)," March 1987, Institute for Interconnecting and Packaging Electronic Circuits, Lincolnwood, IL 60646.

# 6

# Subassembly Interconnection Systems

The use of faster logic switching speeds and higher levels of integrated circuit functional integration are placing new demands on digital system interconnections. At the subassembly level, such equipment is typically designed around the use of a set of modular circuit component subassemblies for optimum end product development, fabrication, assembly, testing and maintenance purposes. [1]

These subassemblies are commonly interconnected through the use of backplanes, printed wiring, connectors, discrete wiring, and combinations of these elements. The selection of such subassembly interconnection configurations must be addressed at the system design level. This is because it is at the system level that the determination is made as to whether data exchange through the subsystem is serial or parallel, what communication protocols are applicable, and thus, what is to be the overall electronic packaging concept.

To further complicate matters, it is also necessary for the subassembly interconnection system to satisfy requirements for increased component density and improved electrical performance. In view of the potential trade-offs between electrical characteristics and other important factors, such as input/output wiring, reliability, repairability, and cost, it is important to quantify the cost-effectiveness and performance of the possible packaging approaches to be used. In this way various designs can be compared in a meaningful manner so that the best implementation hardware can be chosen.

These are just some factors that interconnection systems packaging engineers should consider. In many instances the interconnection system can also actually determine or limit end product performance. Therefore, thorough, up-front packaging of the subassembly interconnection system, with all of the associated electrical and mechanical issues addressed, will help to maximize performance while achieving optimum cost-effectiveness.

## 6.1  BACKPLANE STRUCTURES [2]

Significant increases in packaging density and complexity have led to corresponding changes in backplane subassembly interconnection system design. As the critical link between the plug-in circuit component subassemblies (daughterboards) and the other elements of the end product equipment, backplanes must meet several mechanical and electrical performance requirements.

In its simplest form a backplane (motherboard) is an array of daughterboard-interface connectors that are mounted on a printed wiring board, Figure 6.1, or aluminum panel. The electrical interconnection wiring among the connectors can be provided by the printed wiring board, supplementary discrete point-to-point wiring, or a combination.

The universal trend toward microcomputer-based equipment with bus systems that are fully defined and standardized with respect to their mechanical and electrical parameters has allowed backplanes (and their enclosures) to evolve from purely custom assemblies to semicustom products. [4]

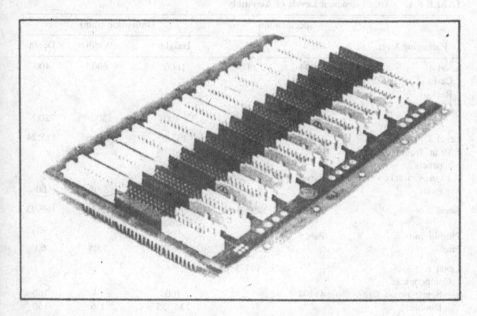

**FIGURE 6.1.**    VME-bus printed wiring board backplane assembly. (*Courtesy of Litton Systems.*)

Perhaps the most influential development has been the growing acceptance of the ''Eurocard'' system as described in Deutsche Industrie Normenausschuss (DIN) specification 41494 and the related International Electrotechnical Commission (IEC) document 279-3. The real benefit of such specifications is that they interrelate the level of equipment packaging, Table 6.1, to enclosure and plug-in subassembly size and the size, type, and quantity of the mating backplane (DIN 41-612) two-part connectors.

## 6.1.1  Aluminum Backplanes

The first backplanes were simple single-layer aluminum panels that mounted the plug-in assembly interface connectors prior to the addition of discrete point-to-point interconnection wiring. Because of its stiffness and planarity, the aluminum panel was primarily used in conjunction with solderless (wire) wrap discrete wiring that was attached to the connector wiring posts either manually, semi-automatically, or by the use of fully-automatic computer-controlled machines.

**TABLE 6.1.    DIN Eurocard Levels of Assembly**

| Packaging level | Specification | | Dimension (mm) | | |
|---|---|---|---|---|---|
| | DIN | IEC | Height | Width | Depth |
| Level 4 | 41494 | 297 | 1600 | 600 | 400 |
| Cases | Parts 1, 3 | | | | |
| Racks | Part 1 | | | | |
| Cabinets | Part 7 | | | | |
| pitch | | | 200 | 300 | 200 |
| Level 3 | 41494 | 297 | 132.5 | 482.6 | 175.24 |
| 19 in. front, back panels | Part 1 | | | | |
| 19 in. subracks | Part 5 | | | | |
| Pitch | | | 44.45 | 5.08 | 60 |
| Level 2 | 41494 | 297-3 | 100 (for single Eurocard) | 15.24 | 169.93 |
| Plug-in units | Part 5 | | | | |
| Pitch | | | 44.45 | 5.08 | 60 |
| Level 1 | | 297-3 | | | |
| Components: | | | | | |
| Single Board Size | 41494 | | 100 | — | 160 |
| Double: | | | 233, 35 | 1.6 | 160 |
| Grids | Part 1 | 97 | | | |
| Holes | 40801, | 97 | | | |
| Pitch | Part 2 | | 44.45 | — | 60 |

In most instances appropriate connector contacts were mechanically attached to the backplane that served as the system ground. Discrete bus bar strips were also used to facilitate power distribution.

The single aluminum backplane eventually gave way to the development of multiple-level aluminum panels and the use of printed wiring boards. The multiple-levels of the aluminum backplanes were insulated from one another so that they could be used for the distribution of the required number of ground and voltage planes. Such backplanes are generally used in products that require a high level of structural rigidity, such as in some airborne, shipboard, and ground-based military applications.

## 6.1.2  Printed Wiring Board Backplanes [5, 6]

The use of printed wiring board backplanes addresses two very prominent concerns, i.e., signal integrity and packaging density. The reasons for this relate to the inherent capabilities of this type of packaging configuration as compared to the use of other interconnection system arrangements.

First, printed wiring is electrically very reliable and dimensionally very stable. Thus it can be used instead of discrete wiring that is dependent to a great degree on the abilities of the manual wiring operator or assembly equipment.

Discrete wiring can often move in relation to other wires such that electrical performance can vary from assembly to assembly or during the lifetime of the product. This is not a problem with printed wiring as the physical relationship of the wiring is controlled through the use of artwork controlled imaging procedures and dimensionally stable materials.

Second, the physical space required for a printed-wiring backplane assembly is significantly less than for discrete-wired interconnection systems. This is because the state-of-the-art for printed wiring fabrication is such that the available space can be used very efficiently.

Lastly, the use of printed-wiring backplanes is cost-effectively competitive with other approaches on an end product basis. This is because of the minimization of hands-on assembly labor content and the broad base of available printed wiring backplane manufacturers. In fact, in many applications the use of printed-wiring backplanes with supplementary discrete wiring provides many users with the optimum benefits of both techniques with respect to design flexibility and signal integrity.

### 6.1.2.1  Standard Printed Board Backplane
### Configurations

The early 1980's gave rise to the establishment of standard backplane busing structures in order to help minimize the recurring engineering costs associated with the design and fabrication of high-performance printed wiring board backplanes. These include the "VME-bus" (Versa Module Electronic) and "Multibus" configurations.

The use of standard backplane configurations such as these offer the packaging engineer many benefits. Including among these is the off-the-shelf availability from multiple vendors for most standardized backplane types, with associated quick delivery and competitive pricing.

### 6.1.2.2  Signal Integrity

The printed wiring board backplane can cost-effectively provide the electrical characteristics that are typically associated with high-performance subassembly interconnection systems. When developing such systems, the packaging engineer should be at least concerned with signal integrity from the source to the destination, and with power distribution quality.

In order to provide the signal integrity needed for modern electronic equipment packaging requirements most printed wiring board backplanes have a multilayer construction that typically provides the necessary ground and voltage distribution planes and establish a controlled characteristic impedance for the signal lines. The signal buses in high-performance backplanes such as these are generally configured in variations of two basic transmission line geometries, Figure 6.2. One configuration is called a microstrip and the other a stripline. [7]

The use of a microstrip or stripline arrangement reduces reflections, reduces crosstalk among the signal lines by providing the appropriate ground returns for the electromagnetic force lines, and increases noise immunity by creating a stable ground reference.

The crosstalk levels can vary greatly due to different backplane designs. It can be reduced by using more layers and increasing the spacing between the signal layers and their associated ground references. Also, since the amount of crosstalk is directly proportional to the rate of change of current versus time, signals with a short rise time and a large surge of current will cause the greatest amount of crosstalk. Thus, when the circuits are operating close to their noise-margin thresholds with high-speed logic, crosstalk minimization becomes a critical concern.

### 6.1.2.3  Power Distribution Quality

Variations in backplane power quality come from four segments of the power distribution network: the power supply, interconnection wiring/cabling, the plug-in functional subassemblies, and the backplane. If the backplane's power-distribution scheme is poorly designed, it can easily be the weakest link in the subassembly interconnection system. Thus, a multilayer printed wiring backplane almost always has at least one full ground plane layer and one wide voltage bus, if not an entire plane. Anything less would certainly be suspect in terms of power quality.

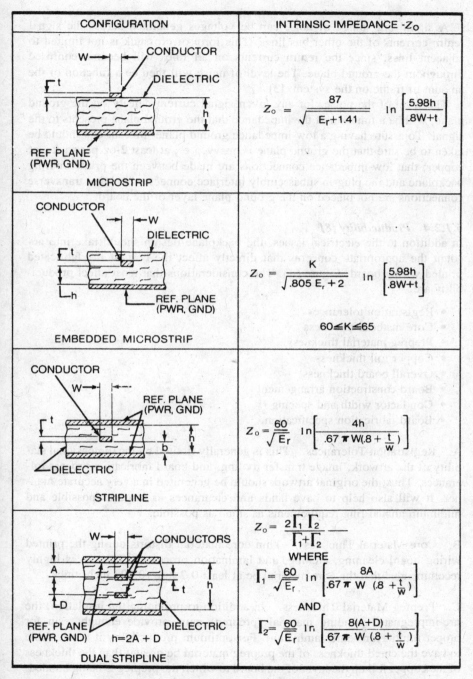

| CONFIGURATION | INTRINSIC IMPEDANCE -$Z_O$ |
|---|---|
| **MICROSTRIP**<br>W → CONDUCTOR<br>DIELECTRIC<br>t, h<br>REF PLANE (PWR, GND) | $$Z_o = \frac{87}{\sqrt{E_r + 1.41}} \ln\left[\frac{5.98h}{.8W + t}\right]$$ |
| **EMBEDDED MICROSTRIP**<br>CONDUCTOR<br>W<br>DIELECTRIC<br>t, h<br>REF. PLANE (PWR, GND) | $$Z_o = \frac{K}{\sqrt{.805 E_r + 2}} \ln\left[\frac{5.98h}{.8W + t}\right]$$<br><br>$60 \leq K \leq 65$ |
| **STRIPLINE**<br>CONDUCTOR<br>W<br>REF. PLANE (PWR, GND)<br>t, h, b<br>DIELECTRIC | $$Z_o = \frac{60}{\sqrt{E_r}} \ln\left[\frac{4h}{.67\,\pi\,W(.8 + \frac{t}{w})}\right]$$ |
| **DUAL STRIPLINE**<br>W<br>CONDUCTORS<br>A, t, h<br>D<br>REF. PLANE (PWR, GND)<br>DIELECTRIC<br>h = 2A + D | $$Z_o = \frac{2\Gamma_1 \Gamma_2}{\Gamma_1 + \Gamma_2}$$<br>WHERE<br>$$\Gamma_1 = \frac{60}{\sqrt{E_r}} \ln\left[\frac{8A}{.67\,\pi\,W\,(.8 + \frac{t}{w})}\right]$$<br>AND<br>$$\Gamma_2 = \frac{60}{\sqrt{E_r}} \ln\left[\frac{8(A+D)}{.67\,\pi\,W\,(.8 + \frac{t}{w})}\right]$$ |

**FIGURE 6.2.** Controlled-impedance printed wiring board transmission line configurations. [7]

123

A significant source of noise can be voltages generated across the signal return currents of the other bus lines. This form of crosstalk is not limited to adjacent lines, since the return currents on all lines will become confused throughout the ground plane. The level of noise will then be a function of the amount of traffic on the system. [3]

The level of the noise, for any given signal current density on the ground plane, will be a function of the impedance that the ground plane presents to the signal. To assure having a low-impedance ground plane, precautions should be taken to be sure that the ground plane is heavy, i.e., at least 2-oz.per sq.ft., of copper; that low-impedance connections are made between the printed wiring backplane and the plug-in subassembly interface connector; and that transverse connections are not placed on the ground-plane layer of the board.

### 6.1.2.4 Producibility [8]

In addition to the electrical issues, the backplane design should take into account the appropriate concerns that directly affect the cost of the fabricated printed wiring board, Figure 6.3. The considerations that most affect producibility are:

- Registration tolerances
- Core material thickness
- Prepreg material thickness
- Copper foil thickness
- Overall board thickness
- Board construction arrangement
- Conductor width and spacing
- Board fabrication specification.

A.   Registration Tolerances.   This is generally limited by the dimensional stability of the artwork, image transfer tooling, and board fabrication process tolerances. Thus, the original artwork should be generated in a very accurate manner. It will also help to have lands and clearances as large as possible and minimum annual ring requirements as small as possible.

B.   Core Material Thickness.   Thin core material distorts during the printed wiring board cleaning, etching, and lamination processes. Thus, it is highly recommended that the core material be at least 0.2 mm (0.008 inch) thick.

C.   Prepreg Material Thickness.   In addition to providing the insulation, the pre-impregnated bonding material (prepreg) must provide enough resin for proper multilayer board lamination. For optimum producibility it is advisable to have the cured thickness of the prepreg material be greater than the thickness of the copper foil on the layer of the board to which the prepreg is being bonded.

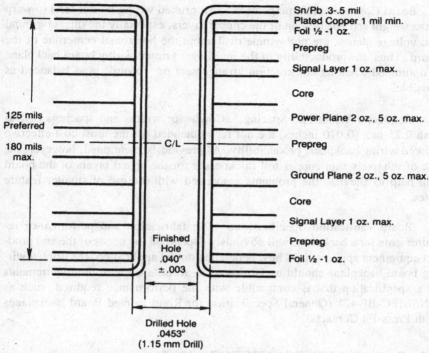

Sn/Pb .3-.5 mil
Plated Copper 1 mil min.
Foil ½ -1 oz.

Prepreg

Signal Layer 1 oz. max.

Core

Power Plane 2 oz., 5 oz. max.

Prepreg

Ground Plane 2 oz., 5 oz. max.

Core

Signal Layer 1 oz. max.

Prepreg

Foil ½ -1 oz.

125 mils
Preferred

180 mils
max.

C/L

Finished
Hole
.040"
± .003

Drilled Hole
.0453"
(1.15 mm Drill)

**Note:** (1) Construction must be symmetric about center line.
**Note:** (2) Core should be used between signal & plane in controlled impedance
applications because of more precise dielectric thickness control.

**FIGURE 6.3.**    Typical balanced multilayer printed wiring board backplane construction. [8]

D.   Copper Foil Thickness.   The use of the thinner copper foils results in more accurate imaging of the board as it minimizes the undercutting of the etched conductors. However, their use also minimizes the amount of copper exposed for through-hole plating. Therefore, thin copper foils (e.g., 0.5 – oz. per sq. ft.) should be used on the outer layers of the multilayer board; internal signal layers should be thicker (e.g., 1-oz. per sq. ft.); and the power and, especially for electrical purposes, the ground and voltage planes should be the thickest (e.g., 2-oz. per sq. ft.).

E.   Overall Board Thickness.   The generally recognized limit on the thickness of multilayer printed wiring board backplanes is four times the diameter of the smallest drilled hole. With press-fit connector termination technology, this limits the overall board thickness to about 4.5 mm (0.180 inches), with the preferred maximum value being 3.2 mm (0.125 inches) for optimum producibility.

F.  Board Construction Arrangement.   All printed wiring boards tend to warp if the weight and positioning of the copper layers, especially the thicker ground and voltage planes, are not symmetrical about the horizontal centerline of the board. Thus, the producibility of the multilayer printed wiring board backplane is optimized when its construction arrangement or "layup" is as balanced as possible.

G.  Conductor Width and Spacing.   Conductor widths and spacings of less than 0.25 mm (0.010 inches) are not recommended for the most cost-effective printed wiring backplane producibility. As previously mentioned, however, the use of relatively thin copper foil thicknesses for the signal layers of the board will help to alleviate the problems associated with the use of smaller feature sizes.

H.  Board Fabrication Specification.   The fabrication and performance requirements for a backplane will obviously depend on the needs of the end product equipment application. Thus, depending on the application, the printed wiring board backplane should be fabricated in accordance with the requirements of a specification that is compatible with the performance required, such as ANSI/IPC-BP-421 (General Specification for Rigid Printed Board Backplanes with Press-Fit Contacts).

## 6.2  BACKPLANE CONNECTORS [2, 9]

Regardless of the type of backplane structure, the connector system can be a determining factor for the performance of the subassembly interconnection system (see Section 2.2.1). The number of contacts needed for each plug-in subassembly, the required printed-wiring board space, the desired level of reliability, and the connector-to-motherboard and discrete wiring options are factors that affect the selection of a backplane connector type and size.

### 6.2.1  One-Part Backplane Connectors

The traditional, most familiar type of backplane connector is the edge-board type, Figure 2.6. This is categorized as a one-part design where plated conductor tabs on the edge of the printed wiring daughterboard mate with the contacts of the motherboard connector.

By nature, though, the edge-board connector arrangement is limited in interconnection capacity by the size of the board edge and the spacing between contacts. Because the printed wiring board acts as the male half of the interconnection system, the insertion and withdrawal forces can be unacceptably high if the board is bowed and/or if the board edge is too long. Also, the normal variations in printed wiring board thickness can significantly affect the mating and reliability of the interconnection.

## 6.2.2    Two-Part Backplane Connectors

The two-part connector, Figure 2.7, with separate male and female contacts, have gained in popularity for use with printed wiring backplanes in order to maximize interconnection density and maintain control over mating characteristics. This type of connector is available in a wide variety of types and sizes to accommodate a wide variety of end product applications.

### 6.2.2.1    Blade-and-Fork Contact Connectors

Two-part connector with male, plug-in "blade" and female, backplane "tuning fork" contacts are suitable for use in high-reliability applications. The basic design was originally developed for use with aluminum backplanes, although they are now more commonly used with printed-wiring board backplanes. Both through-hole and surface mount high-density types are available.

### 6.2.2.2    Post-and-Box Contact Connectors

The use of two-part connectors with male "post" and female "box" contacts has been very popular, especially in DIN standard applications. The original DIN connectors had the conventional arrangement of the male contacts on the daughterboard and the female contacts on the motherboard. However, the suitability of the 25-mil square post for terminating supplementary discrete backplane solderless-wrap wiring has advanced the use of an "Inverse DIN" configuration, Figure 2.7, with the posts serving the multiple purpose of mating with the plug-in assembly, terminating to the printed wiring backplane, and being the terminal for the discrete wiring.

The DIN system of connectors is well suited for applications that require a greater contact density than is feasible with the edge-board types. Each DIN connector housing can be selectively loaded with up to 96-contacts on a three-row 2.54 × 2.54 mm (0.100 × 0.100 inch) grid.

Derivatives of the VME-bus application exist in which as many as three 96-contact DIN connectors are used with each plug-in subassembly. However in these instances drawbacks such as accumulated tolerances and board bowing must be taken into account.

## 6.2.3    High-Density Contact Connectors

Special backplane connector designs provide a greater degree of interconnection density than do the conventional blade-and-fork or post-and-box types while adding other refinements, such as the use of low insertion force (LIF) contacts to facilitate connector mating. Unlike the standard DIN connectors, these high-density connectors differ from manufacturer to manufacturer and can have up to five rows of contacts on centers as close as 1.27 (0.050 inch) or less within a row. Such interconnection systems can accommodate up to 800 or more contacts per plug-in subassembly.

### 6.2.4    Mechanical Considerations

Beyond the obvious advantages of increased interconnection density, two-part connector systems provide more control of mating and unmating characteristics, such as insertion/withdrawal forces, and normal contact forces. As contact density (and quantity) increases these important parameters can be controlled by the connector manufacturer since they are not subject to the dimensional variances of the daughterboard.

#### 6.2.4.1    Contact Forces
In order to implement a high-density application, the insertion forces between mating contacts should be minimized to about 2 ounces or less per mated contact-pair. However, as insertion forces are reduced there is a corresponding reduction in contact normal force. Since this force is a good indication of the reliability of the interconnection, it is important that the normal force be maintained at a sufficiently high level of at least 85 grams.

In order to minimize the insertion/withdrawal forces and maximize the contact engagement normal force, such LIF connectors employ special features that take into account the contact material, geometry, and plating.

#### 6.2.4.2    Alignment
Precise alignment between mating connectors and contacts is vital if high-density backplane interconnection systems are to work properly. It is usually necessary to adopt the use of a systems approach to backplane design, fabrication and assembly to achieve the required level of mating accuracy. Such systems approaches take into account the relationships among the various elements of the interconnection system in order to make sure that the finished assembly cost-effectively satisfies performance (reliability) requirements.

Accurate alignment is clearly a function of the relationship among these system-level components as shown in Figure 6.4. The first level of hardware interface is between the plug-in subassembly and the card guide that initially positions it prior to connector mating with respect to the backplane enclosure. The other critical interface locations are between the connector insulator (header) housings, between the daughterboard header and its contacts, and finally, between the mating contacts. At each of these interface locations specially designed lead-ins should exist to provide the degree of alignment that is necessary

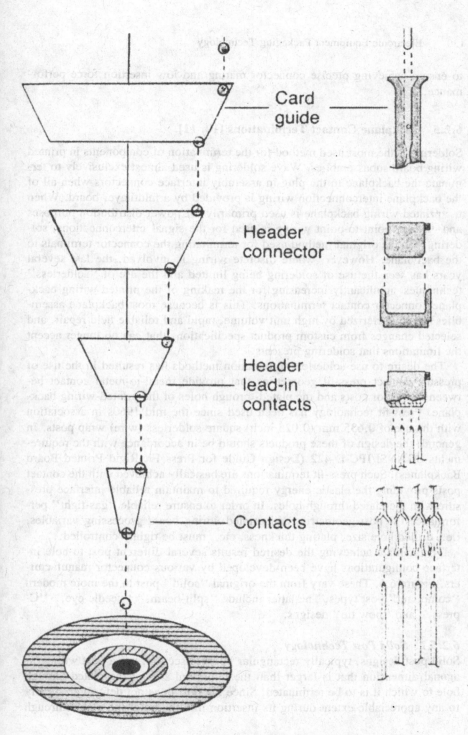

**FIGURE 6.4.** Plug-in subassembly and backplane alignment. [2]

to ensure achieving precise connector mating and low insertion force performance.

### 6.2.5  Backplane Contact Terminations [10, 11]

Soldering is the most used method for the termination of components in printed wiring board subassemblies. Wave soldering is used almost exclusively to terminate the backplane to the plug-in assembly interface connectors when all of 'he backplane interconnection wiring is provided by a multilayer board. When th. ' printed wiring backplane is used primarily for power distribution purposes ana 'iscrete point-to-point wiring is used for the signal interconnections, soldering was the original method used for terminating the connector terminals to the bach plane. However, where discrete wiring is involved, the last several years has seen the use of soldering being limited and the use of "solderless" techniques significantly increasing for the making of the printed-wiring backplane connector contact terminations. This is because most backplane assemblies are characterized by high unit volume, rapid and reliable field repair, and selected changes from custom product specifications that can no longer accept the limitations that soldering presents.

The desire to use solderless termination methods has resulted in the use of pressure-contact press-fit geometries that provide metal-to-metal contact between connector posts and the plated-through holes of the printed-wiring backplane. Press-fit technology has been used since the mid 1960s in association with the use of 0.635 mm (0.025 inch) square solderless (wire) wrap posts. In general, the design of these products should be in accordance with the requirements of ANSI/IPC-D-422 (Design Guide for Press Fit Rigid Printed Board Backplanes). Such press-fit terminations are basically achieved with the contact posts providing the elastic energy required to maintain reliable interface pressure with the plated-through holes. In order to ensure reliable "gas-tight" performance the post geometries and printed wiring board processing variables, i.e., drilled hole size, plating thickness, etc., must be tightly controlled.

To facilitate achieving the desired results several different post-to-hole interface configurations have been developed by various connector manufacturers, Figure 6.5. These vary from the original "solid" post to the more modern "compliant" post types. The latter include "split-beam," "needle eye," "C-press," and "bow tie" designs.

#### 6.2.5.1  Solid Post Technology
Solid post designs, typically rectangular in cross-section, use a post with a diagonal dimension that is larger than the minimum size of the plated-through hole to which it is to be terminated. Since the post does not deform elastically to any appreciable extent during its insertion into the hole, the plated-through

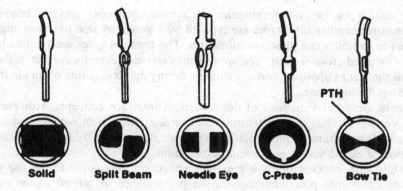

**FIGURE 6.5.**    Various press-fit post configurations. [10]

hole must expand or otherwise deform to accommodate the interference fit. Consequently, all of the termination energy is stored in the printed wiring backplane.

Most of this force is transformed into permanent board deformation with only a small part of it converted to stored energy. Thus, depending upon backplane thickness and the degree of engagement interference, board damage may occur, as may low post retention forces. Therefore, although less costly to produce, the use of solid press-fit posts with printed wiring backplanes have a number of limitations when compared to the compliant types, such as:

• Higher post insertion forces and, thus, potentially more post buckling
• Potentially more printed wiring board deformation and damage
• More localized plated-through hole stresses and plating damage
• Tighter tolerances must be maintained
• Greater incident of post loosening due to the board's inability to store elastic energy
• Repairability may require the use of over-sized replacement posts or undesirable soldering.

### 6.2.5.2  Compliant Post Technology

The emergence of "compliant" press-fit designs was brought about by the need to overcome the limitations of solid-post technology. In these instances, the interface between the contact post and the wall of the plated-through hole is based primarily on the designed-in deformation of the post rather than of the hole. The degree of compliance is determined by the ability of the post to deform elastically, thus storing elastic energy to optimize interconnection performance.

The plated-through hole interface portion of compliant post designs are of three basic types, i.e., non-rectangular solids, crescent shapes, and split beams. The non-rectangular solid types are typified by a geometry that has been engineered to exhibit some plastic deformation. The crescent types are typified by their C-shaped cross-section. The split-beam versions generally consist of two beams that act as opposing spring members during the flexing that occurs in the insertion flexing mode.

There are other variations of the solderless interface concepts. However, each version exhibits unique deformation characteristics, both radially and axially, with the ultimate connection performance determined by the degree and location of elastic versus permanent deformation.

While each concept offers a press-fit connection alternative to the use of conventional solid posts, interconnection performance as related to the end product application varies significantly from one post type to another.

## 6.3  DISCRETE SOLDERLESS-WRAP BACKPLANE WIRING [12, 13]

Subassembly interconnection system technology has been continually evolving since the early days of discrete hand-soldered point-to-point interconnections which were used between solder posts on the first one-part printed wiring connectors. The impetus toward developing improved wiring methods was supplied by the introduction of the use of solderless (wire) wrapping processes by Bell Laboratories in the early 1950s.

Most electronic equipment subassembly interconnection systems use multilayer printed wiring exclusively, or hybrid combinations of printed wiring and aluminum backplane structures with discrete solderless (wire) wrapping technology to form various cost-effective and reliable end product variations. However, as described in ANSI/IPC-DW-425 (Design and End Product Requirements for Discrete Wiring Boards), other methods are used to satisfy special discrete wiring applications.

The choice of a particular combination (or the exclusive use of only one approach) depends on the application. Some factors to consider are:

- Physical space
- Circuit operating speed
- Power distribution
- Design flexibility
- Maintainability
- Repairability
- Production quantity
- Cost effectiveness.

### 6.3.1 Solderless Wrapping Versus Soldering

Solderless (wire) wrapping is a process for connecting solid-conductor discrete wires to a terminal by tightly coiling the wire around the terminal wire with a hand tool or machine, Figure 6.6. When compared to hand soldering, the advantages of using this solderless wire termination method are clear. For example:

- Solderless (wire) wrapping does not require the use of heat to make the termination. This, eliminates the possibility of component damage and improves operator safety.

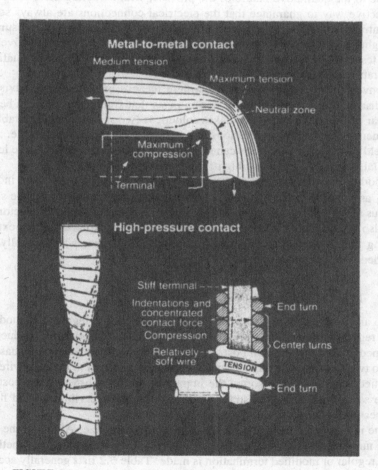

**FIGURE 6.6.**     Basic elements of solderless (wire) wrap terminations. [12]

- Solderless (wire) wrapping allows for compact terminal spacing, i.e., 2.54 mm (0.100 inch), that requires significant operator skill when terminated with hand soldering.
- Solderless (wire) wrapping can be used to cost-effectively terminate discrete wires through the use of either manual, semi-automatic, or fully automatic methods.
- Perhaps most importantly, solderless (wire) wrapping can produce reliable terminations. Due to the controlled use of precision tooling, such terminations can be made consistently regardless of operator skill levels.

Due to the qualitative nature of the process, with soldering there is no nondestructive way to guarantee that the electrical connections are always sound. In contrast, solderless (wire) wrapping provides a quantitatively measurable, gas-tight, cool termination with a design life of 40 years. (Nondestructive strip force tests can be used to provide a reasonable determination of the quality of the wrap-wired termination.)

Conversely there are applications where the use of soldering is advisable. One disadvantage, is the need for wiring posts that significantly extend beyond the rear of the backplane. This makes solderless (wire) wrapping unsuitable for high-density packaging applications that require a low backplane profile. Also, the field of wrapped-wire posts can significantly increase crosstalk/noise levels, especially for high-speed and high-frequency applications.

Another drawback is that each end of the wire must be terminated individually, as opposed to the use of a mass soldering process that can make simultaneous terminations. Solderless (wire) wrapping with only solid-conductor wire can also be unsuitable for connections between the backplane and the external cabling that is subject to bending and flexing, and thus, which usually uses stranded wire for this purpose.

## 6.3.2  Types of Solderless (Wire) Wrapping

There are two basic types of wrapped-wire connections: regular and modified. In a regular termination, only the bare uninsulated solid-wire conductor is wrapped around the terminal post, Figure 6.6. A modified termination has from one to two and one-half turns of insulated wire in addition to the bare wire. The modified wrap is used to soften the impact of the first corner of the post, and at the same time, serves as strain relief against possible vibration and flexing stresses on the wire.

The number of wire turns is a function of the wire size. Although the minimum number of turns is a function of individual requirements and whether or not a regular or modified termination is made. Table 6.2 lists generally accepted values.

**TABLE 6.2.     Solderless Wrap Wire-Turn Parameters [12]**

| Wire size | Diameter | | Minimum number of turns | |
|---|---|---|---|---|
| | Inches | MM | Modified | Regular |
| 30[a] | 0.0100 | 0.25 | 7 stripped plus 1/2 insulated | 7 stripped |
| 28[a] | 0.0126 | 0.32 | 7 stripped plus 1/2 insulated | 7 stripped |
| 26 | 0.0159 | 0.40 | 6 stripped plus 1/2 insulated | 6 stripped |
| 24 | 0.0201 | 0.51 | 5 stripped plus 1/2 insulated | 5 stripped |
| 22[b] | 0.0253 | 0.64 | 5 stripped plus 1/2 insulated | 5 stripped |
| 20[b] | 0.0320 | 0.81 | 4 stripped plus 1/2 insulated | 4 stripped |
| 18[b] | 0.0403 | 1.02 | 4 stripped plus 1/2 insulated | 4 stripped |

[a]Not recommended for 0.45 × 0.45 and 0.30 × 0.60 wrapposts.
[b]Not recommended for 0.25 × 0.25 wrapposts.

## 6.3.3   Solderless Wiring Terminals

All solderless (wire) wrap terminals (posts) have at least two diagonally opposed sharp edges to which the gastight connection is made. In most applications the posts are square in cross-section, i.e. 0.635 mm (0.025 inch) for small wires and 1.15 mm (0.045 inch) for large wires, Table 6.3.

The exposed wrappable length of the post is a function of the type of wrap, wire size, and the number of levels of terminations to be made. When 24- and 26-AWG wire and three levels of modified termination are used the extended post length should be a minimum of 19.0 mm (0.750 inches) and a maximum of 22.2 mm (0.875 inches); for the smaller 28- and 30-AWG wires these values can be reduced to 1.3 mm (0.500 inches) and 1.6 (0.625 inches), respectively.

The metallurgy of the wrapping post depends, obviously, on the requirements for the other end of the terminal. Thus for backplane connector applications, the post base material is generally either beryllium copper or phosphor bronze. Gold plating is used for optimum reliability. In some instances, the wire termination end of the post is also tin plated or solder coated when wiring changes or field repairs will be made by soldering.

## 6.3.4   Discrete Wrapping Wire

The size of the wire used is usually dependent on the application. In telecommunications applications this is typically in the 22- to 24-AWG range; for most

**TABLE 6.3.    Solderless (Wire) Wrap Post Parameters [13]**

| Wire size | Wrap post size | Wrap post grid |
|---|---|---|
| 24 Gage | 0.045″ × .045″ (1,14 mm × 1,14 mm)<br>0.031″ × .062″ (0,79 mm × 1,57 mm) | 0.156″ × 0.200″ (3,96 mm × 5,08 mm)<br>  Rectangular<br>0.200″ (5,08 mm) Square<br>0.250″ (6,35 mm) Square<br>0.250″ (6,35 mm) Staggered |
| 26 Gage | 0.045″ × .045″ (1,14 mm × 1,14 mm)<br>0.025″ × .025″ (0,64 mm × 0,64 mm) | 0.156″ × 0.100″ (3,96 mm × 2,54 mm)<br>  Rectangular<br>0.150″ (3,81 mm) Square<br>0.125″ (3,18 mm) Square<br>0.150″ (3,81 mm) Staggered<br>0.100″ (2,54 mm) Staggered |
| 30 Gage | 0.025″ × .025″ (0,64 mm × 0,64 mm)<br>0.020″ × .030″ (0,51 mm × 0,76 mm) | 0.156″ × 0.100″ (3,96 mm × 2,54 mm)<br>  Rectangular<br>0.156″ × 0.200″ (3,96 mm × 5,08 mm)<br>  Rectangular<br>0.100″ (2,54 mm) Square<br>0.125″ (3,18 mm) Square<br>0.150″ (3,81 mm) Square<br>0.200″ (5,08 mm) Square<br>0.250″ (6,35 mm) Square<br>0.250″ (6,35 mm) Staggered<br>0.150″ (3,81 mm) Staggered<br>0.100″ (2,54 mm) Staggered |

other electronic applications it is from 26- to 30-AWG. Power buses may require the use of heavier gauge wire.

Oxygen-Free High-Conductivity (OFHC) copper is the most widely used base metal for electronic applications. However, tough-pitch copper, certain copper alloys, and copper-steel conductors are also used.

An additional tin or silver plating can be applied for extra corrosion resistance. However, a properly terminated wire provides a gastight metal-to-metal connection between the base conductor and post with a resistance of 0.0001 ohms, so that the advantage of using plating is lost once the wire is wrapped. The wire insulation is typically polyvinyl chloride (PVC), polyester, or polyvinylidene fluoride (PVDF) which has a low adherence to metal, and thus is easily stripped. For high-temperature and high-vibration environments or where abrasive resistance is needed, other insulating materials, such as polyimide H-film or fluoropolymers (TFE or FEP Teflon), may be used, although some of these are generally not as strippable. The final selection of an insulation material is based on its insulation properties, cut-through resistance, price, automatic machine suitability (if applicable), etc., Table 6.4.

**TABLE 6.4.** Typical Solderless Wrap Wire Insulation (*Courtesy of Brand-Rex Co.*)

| Characteristics | PUDF | TFE Teflon | FEP Teflon/Nylon | Polysulfone | PVC/Polyester | Irradiated PVC | Semi-rigid PVC | PVC/nylon | FEP Teflon/polyimide |
|---|---|---|---|---|---|---|---|---|---|
| Dielectric constant 1 MHz | 6.4 Nom. | 2.6 Nom. | 2.2 Nom. | 3.1 Nom. | 3.5 Max. | 2.7 Nom. | 4–5 | 4.0 Nom. | 2.2 Max. |
| Cut-through resistance | Good | Excellent | Good | Good | Excellent | Excellent | Fair | Fair | Excellent |
| Stripping consistency | Excellent | Excellent | Good | Excellent | Good | Good | Fair | Fair | Fair |
| Minimum curl | Good | Excellent | Good | Excellent | Fair | Excellent | Excellent | Good | Good |
| Shrink-back resistance | Good | Excellent | Good | Excellent | Fair | Excellent | Good | Fair | Good |
| Temperature range, °C | −65 to +130 | −80 to +150 | −55 to +115 | −65 to +130 | −55 to +105 | −55 to +105 | −55 to +80 | −55 to +105 | −80 to +200 |
| Concentricity consistency | Excellent | Excellent | Good | Excellent | Excellent | Good | Good | Fair | Good |
| Curl resistance | Good | Excellent | Good | Excellent | Fair | Excellent | Excellent | Good | Good |
| Chemical resistance | Good | Excellent | Good | Poor | Fair | Excellent | Good | Fair | Excellent |
| Cost | Moderate | High | High | High | Very High | Low | Low | Low | Very High |
| Termination method | Automatic | Automatic | Manual | Automatic | Automatic | All methods | Manual | Manual | Automatic |
| Conductor compatibility | Silver-plated | Silver-plated | Silver-plated | Silver or tin-plated | Silver or tin-plated | Silver or tin-plated | Silver or tin-plated | Silver or tin-plated | Silver-plated |
| Packaging density | Low | High | High | Low | Medium | Medium | Low | Low | High |

### 6.3.5  Wiring Grids

For optimum wire routability, solderless (wire) wrap backplanes have the terminal posts located on one of the recommended grid configurations, Table 6.5. Such grids are generally either in a rectangular or staggered pattern.

The rectangular (or square) grid system has rows of terminals that coincide with backplane grid lines, with a terminal located at each basic grid intersection. The staggered grid system has alternate rows of terminals that are offset in one direction.

**TABLE 6.5.    Solderless (Wire) Wrap Processing Considerations [14]**

| Question | Impact |
|---|---|
| What type of insulation is used on wire? | Exotic insulations may impact the per wire cost due to higher raw material cost. |
| How large is the panel? | In addition to possible physical machine limitations, it is important to remember that long wires increase machine movement and may impact the per wire cost. |
| Is a from-to list available? | A from-to list is the most efficient input media for wire wrapping sequence tape generation. It is also a necessity for computer generated circuit analyzer drive tapes. |
| What wire size is to be wrapped? | In general, smaller sizes are more difficult to wrap and may impact the per wire price. |
| Are twisted pairs or special routing present? | Software generation and per wire cost increases are involved. |
| Is a panel available for geometry program generation? | Development of geometry program from a panel gives the highest confidence level and facilitates testing of the program. |
| Is a from-to list available on a punched card deck? | This method of transferring data gives a high confidence level. |
| Does the customer have a wire wrapping specification? | If not, should we use our own internal specification, EIA specification, Military specification, etc.? |
| Does the customer wish to supply his own software? | The tape must be in the computerwrap binary format. If other tape formats are available, it is necessary to check out the feasibility of conversion to computerwrap binary. |
| What routing style is desired? | Routing style affects price per wire. |
| How many wires are on the panel? | Wire count determines computer run charge. |
| If the panel is to be automatically tested, can customer supply mating connectors and/or paddle cards? | Availability of mating connectors from customer will drastically reduce the price of the test interface cable. |
| If panel is to be automatically tested, is it I.C. socket or card cage configuration? | While card cage configurations are easily accessed with adapter cables, integrated circuit socket types must be accessed via spring pin fixtures. |

Each arrangement has distinct advantages over the other. The square grid has the greatest post density in a given area of the backplane. However, this usually limits wire dressing or placement to certain specific channels because of the close terminal spacing.

The staggered grid, with its reduced terminal density, has a greater wire-routing channel width. This is advantageous with circuits where electrical noise and crosstalk problems are critical, as there is less parallel wiring within a given channel.

### 6.3.6  Wiring Equipment

There is a wide variety of solderless (wire) wrapping and unwrapping tools available for backplane applications that vary from simple screwdriver-like hand tools costing as little as $10 to fully automatic systems costing $150,000 or more. However, convenient power driven manual and semi-automatic models are the most widely used and applicable for all but the highest volume discrete wiring purposes. Regardless of which approach is taken there are several basic processing considerations that must be taken into account, including those enumerated in Table 6.6. [14]

There are, obviously, distinct achievable performance levels that depend on the type of equipment used. In general, the wiring rates (wires terminated per hour) are in the area of 10-50 (hand), 200-300 (semiautomatic), and 1,000 (fully automatic); the corresponding error rates are 5–10%, 0.5%, and 0.1%, respectively.

#### 6.3.6.1  Manually Operated (Hand) Wrapping Tools

Hand-wrapping tools are especially useful for small lot production, prototyping, and repair purposes. Unwrapping tools are often of the manually operated type since this operation is typically only done occasionally.

An enhanced type of hand operated equipment is the trigger actuated, manually powered, wrapping tool. However, the use of these tools is not recommended for the termination of a large number of wires, especially large wires, because of the associated wear and operator fatigue factors. When this is the case, battery-, electric- and pneumatic-powered hand tools are commonly used. Battery-powered tools are best for low quantity applications that are not too taxing on the power source, even though rechargeable batteries can be used. The lack of a power cord adds to their portable, quiet, and safe operation.

Electric alternating-current power tools are the most widely used wrapping units. They are typically quieter and lighter than their pneumatic-powered alternatives.

**TABLE 6.6.    Solderless (Wire) Wrap Fully-Automatic Processing Considerations [14]**

| Question | Impact |
| --- | --- |
| Was subject panel initially designed for automatic wrapping? | The parameters required for automatic wrapping cannot be built into a panel after the fact. |
| What is panel size and active wrap area? | Maximum panel size is 26 × 26 inches and maximum active wrap area is 22 × 22 inches. |
| Is the panel fabricated, packaged, and delivered with positioning within a 0.010 inch radius? | Pins may have to be adjusted to specification. |
| Is more than one grid spacing present on the panel? (For example: 0.100 × 0.125 inch). | Since the panel is rotated during the wrapping cycle, pin displacement must be equal in the X and Y directions. This is necessary to employ the routing fingers when the X and Y directions swap during rotation. |
| What is wire size to be applied? | It is an expensive proposition to change a machine to a wire size other than the size that it was built for. |
| What is the tolerance of the outside dimension of the required wire? | The machine requires wire with a dimensional tolerance of ±0.001 inch with 80% concentricity. |
| Is more than one wire size present on the panel? | The machine cannot do more than one wire size at a time. With mixed sizes on the same panel, the remaining gages would have to be done on another automatic machine, a semi-automatic machine, or by hand. |
| Are twisted pairs employed on the panel? | Twisted pairs cannot be applied by the automatic equipment, they would be applied by semi-automatic machines or by hand. |
| What are outside pin dimensions? (For example: 0.025 × 0.025 inch for 30 gage wire.) | Tooling must conform to pin size. |
| Do all pins exhibit the same outside dimensions? | The machine is capable of wrapping only one pin size at a time. |
| What is the tolerance of the pin diagonal measurement? | The wrapping bits will accept a tolerance of no greater than ±0.001 inch. |
| Are all pins the same height? | Only one pin height can be set for a given run. |
| Do any components protrude above the base of the pins within the wrap area? | Any protrusion above the base of the pins will cause the cut and stripped wire to be pulled from the Wire-Wrap bits, since the wire is cut to the exact length to make the wire run and loaded into both bits simultaneously. Wrap height can be raised to clear components as long as the Wire-Wrap pins are long enough to accommodate the required number of wraps. |
| What routing style is required? | The machine is capable of picture frame routing only. |
| Are there any restricted areas on the panel? | Restricted areas will bring about a need for an extensive programming effort. |

### 6.3.6.2  Semiautomatic Wrapping Equipment

The use of semiautomatic solderless (wire) wrapping equipment, Figure 6.7, is basically associated with the use of $X$-$Y$ coordinate locating machines that are use in conjunction with powered hand tools. As already noted, the use of these machines can dramatically increase wiring rates and reduce wiring errors when compared to the use of the hand tools in an unaided manner.

There are two basic types of semiautomatic equipment, i.e., those with a fixed head and those with a moving pointer. With a fixed head the backplane moves with respect to the wrapping tool; with a moving pointer, Figure 6.7, the tool position moves with respect to the backplane.

When used with numerically-controlled systems, the semiautomatic machine reads the encoded location and decision data, locates the wiring point, and provides the operator with the necessary information to perform the wire termination, such as length of wire, whether or not it is the first or second wrap on

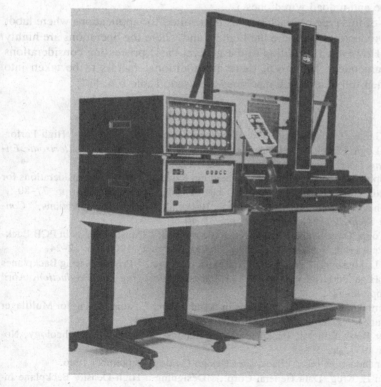

**FIGURE 6.7.**    Typical moving-pointer semiautomatic solderless (wire) wrap equipment. (*Courtesy of OK Industries.*)

the wire, and the wiring sequence number. Unfortunately, there are no controls over the routing configuration of the wire between its end termination points.

Because of various factors, the use of semiautomatic solderless (wire) wrapping machines are an economical choice for a wide range of wiring volumes. In fact, even in high-volume applications, the use of several semiautomatic units is preferred to the use of one fully-automatic machine.

### 6.3.6.3  Fully-Automatic Wrapping Machines

The fully-automatic solderless (wire) wrapping machines perform the entire wiring operation without the aid of human assistance. This includes cutting reel-fed wire to predetermined lengths, stripping both ends of the wire, forming the wire into its routing configurations, positioning the wire and both of the wrap heads simultaneously over desired terminal posts, lowering the wire and wrapping tools to the desired level of the post, and terminating both ends of the wire at one time. (The main role of the operator is to load unwired backplanes into the machine and unload wired ones.)

The use of this type of machinery is best suited for applications where labor costs and production levels are the highest and where the operations are highly repetitive. However, in addition to the general basic processing considerations already enumerated, Table 6.6, there are additional factors to be taken into account when fully-automatic machines are used, Table 6.6. [14]

### References

1. Mark Gailus, Christopher Heard, Teradyne Connection Systems, "High-Performance Logic Places New Demands on Backplane Interconnects," *Electronic Engineering Times*, November 1985, pp. T24–T25.
2. Christopher Van Veen, Teradyne Connection Systems, "Design Considerations for Backplane Interconnection Systems," Electri-Onics, February 1986, pp. 77–80.
3. C. Michael Hayward, Hybricon Corp., "Backplane Design Considerations," *Connection Technology*, October 1986, pp. 35–40.
4. Mike Topp, Dave Sargent, BICC-Vero Electronics, "Considerations in PCB Backpanel Packaging," *Electronic Manufacturing*, February 1990, pp. 22–24.
5. Gerald L. Ginsberg, Component Data Associates Inc., "Printed Wiring Backplanes Reach High Performance Levels," *Electronic Packaging and Production*, April 1985, pp. 48–53.
6. Jim Steranko, Augat Inc., "Choices in Standard and Custom Busing for Multilayer Backplanes," *Electr-Onics*, December 1986, pp. 50–53.
7. Jennifer Rose, Editor, "Backplane Design Guide," Connection Technology, November 1987, pp. 37–39.
8. Bevmar Industries Inc., "Backpanel Design Guide," September 1986.
9. Stephen E. Krug, Data General Corp., "Designing a High-Density Backplane Interconnect System," *Connection Technology*, April 1988, pp. 26–30.

10. Francis Dance, Winchester Electronics, "Reliability Characteristics of Compliant Press-Fit Pin Connectors," *Electronic Manufacturing*, October 1988, pp. 28–32.
11. Paul Gazzara and Frank Siano, Thomas & Betts Corp., "How Pin Geometry Affects Press-Fit Connections," *Connection Technology*, March 1987, pp. 29–32.
12. David Leventhal, OK Industries, "Wire Wrapping Offers Something for Everyone," *Electronic Packaging and Production*, October 1985, pp. 136–141.
13. Universal Instruments Corp., "Wire Termination," Design Guide, Section WT.
14. Anon, "Discrete Wire Terminations," *Assembly Engineering*, July 1983, pp. 10–13.

# 7

# Wiring and Cabling Assemblies

There is no shortcut to the optimum selection and design of wiring and cabling assemblies (harnesses) to interconnect electronic subsystems. However, many established wire, cable, and connector types which are available for this purpose are compatible with the use of reliable assembly (termination) techniques, Table 7.1.

## 7.1 ELECTRICAL CHARACTERISTICS [1]

The selection of the appropriate wiring and cabling assembly components depends to a great extent on the performance requirements for the end-product application. Thus, before the actual wire/cable and connectors are chosen the packaging engineer must have a clear understanding of the electrical, as well as the physical and environmental, considerations that must be taken into account. (See also Section 5.1.)

The electrical considerations can be extremely complicated, since they involve to varying degrees voltages, current carrying capacity, circuit speed and frequency, attenuation, capacitance, velocity of signal propagation, inductance, source and load impedance, corona, and electrostatic and electromagnetic interference, to name a few. These electrical performance requirements generally dictate the configuration of the wire or cable to be used, Table 7.2, and thus, limit the selection of an appropriate connector. The variability of these electrical requirements has led to the availability of a wide variety of interconnection wiring and cabling products.

144

**TABLE 7.1. Wiring and Cabling Assembly Options** (*Courtest of Burndy Corp.*)

| TO BE CONNECTED | CONNECTOR TYPE | TERMINATION METHOD |
|---|---|---|
| Wire to wire, solid or stranded | Rectangular metal shell | Crimp removable contacts |
| | | Solder |
| | | Wire wrap |
| | | Insulation displacement mass termination |
| | Rectangular shelless, self mounting | Crimp removable contacts |
| | | Solder |
| | | Wire wrap |
| | Splice | Crimp |
| Coaxial, shielded or twisted pair to same | Rectangular metal shell | Crimp removable contacts |
| | | Solder |
| | | Insulation displacement mass termination |
| | Rectangular shelless, self mounting | Crimp removable contacts |
| | | Solder |
| Stranded and solid wire, coaxial and shielded wire and twisted pair wire to each other in any combination | Rectangular metal shell | Crimp removable contacts |
| | | Solder |
| | | Insulation displacement mass termination |
| | Rectangular shelless, self mounting | Crimp removable contacts |
| | | Solder |
| Wire to wire, solid or stranded | Round, trilock or threaded coupler type—with an environmental seal | Crimp |
| | | Solder |
| | Round, trilock or threaded coupler type—without an environmental seal | Crimp |
| | | Solder |
| Coaxial, shielded or twisted pair to same | Round, trilock or threaded coupler type—with an environmental seal | Crimp |
| | Round, trilock or threaded coupler type—without an environmental seal | Crimp |
| | | Solder |
| | RF type | Crimp |
| | | Solder |
| Stranded and solid wire, coaxial and shielded wire and twisted pair wire, to each other in any combination | Round, trilock or threaded coupler type—with an environmental seal | Crimp |
| | Round, trilock or threaded coupler type—without an environmental seal | Crimp |
| | | Solder |
| | RF type | Crimp |
| | | Solder |

**TABLE 7.2.    Typical Signal-Transmission Cable Characteristics [5]**

| Characteristic | Cabled twisted pairs | Coaxial Cable | | Tri-lead | Flat cable | Fiber optics |
|---|---|---|---|---|---|---|
| | | Solid core | Air-spaced | | | |
| Impedance tolerance | P | E | E | G | E | — |
| Attenuation | F | E | E | G | G | E |
| Crosstalk | G | E | E | G | F-G | E |
| Time delay | P-F | G | E | G | G | P |
| Rise time | F | G | E | G | G | E |
| Bandwidth | F | G | E | G | F | E |
| Mechanical integrity | E | G | G | F | F | P |
| Flexibility | G | G | G | E | E | P |
| Cable dimensions | P | E | E | G | G | E |
| Dimension tolerance | F | G | G | G | E | E |
| Cable cost | E | G | G | F | F | G |
| Installed cost | F | F | F | F | E | P |

Notes
E—Excellent    G—Good    F—Fair    P—Poor

## 7.1.1  Wiring and Cabling Circuit Speed and Frequency

More care must be given to component and assembly method selection because of the higher insulation losses and the need to match impedances when high circuit speeds and operating frequencies are involved. The combining of high power levels and high operating frequencies can also create additional problems with respect to additional losses and overheating.

## 7.1.2  Wiring and Cabling Voltage Safety Factor

One electrical consideration that must not be overlooked is voltage safety factor, i.e., the insulation breakdown voltage divided by the actual maximum operating voltage. The highest safety factors have to be chosen for multiconductor cables that are subjected to considerable movement (flexing). The high repair and replacement costs of such a cable assembly, plus the large number of potential failure points within it, account for a safety factor of between 70 and 100 for each conductor.

This can be reduced to between 10 and 20 for discrete hookup wire when it is used in fixed applications that are seldom flexed. The lowest voltage safety factor is recommended for high-voltage cables because it is often impractical to make them with higher values. In any case, the safety factor should be higher when the environment is hard on the cabling.

### 7.1.3  Wiring and Cabling Capacitance

With pulsed or alternating current signals the cabling capacitance must be charged before the load can receive full voltage signals from the transmitting source. Sometimes the load voltage may never reach the full source value, and it may have a different wave shape because the interconnection wiring assembly is acting as a filter. Thus, in order to minimize capacitance problems an insulation with a low dielectric constant should be chosen, the conductors should be kept as far apart as possible, and the cabling length should be minimized.

### 7.1.4  Wiring and Cabling Attenuation

Attenuation is a property to consider carefully in transmission line applications with low signal levels. A cable's attenuation is an indication of losses due to generated heat. The conductor is partially responsible for this because of its resistance. So is the alternating current capacitance loss factor that, for two cables of the same dimensions, is proportional to the product of the dielectric constant and the dissipation factor.

Therefore, in order to minimize attenuation it is advisable to follow the same guidelines just mentioned to minimize capacitance. Also, it is important to make sure that the insulation has a low dissipation factor and that, as far as is practical, conductors with the highest conductivity are used.

However, care must be exercised even though the use of larger conductors reduces resistance losses. For example, this generally increases alternating current losses in a coaxial cable of constant diameter. The increased loss stems from the increased capacitive coupling between conductors. Such losses also increase at higher circuit operating frequencies.

### 7.1.5  Wiring and Cabling Velocity of Signal Propagation

Another important electrical characteristic is the velocity of signal propagation. This factor is inversely proportional to the insulation dielectric constant for similar cabling configurations. Therefore, the term is often conveniently used to indicate the relative excellence of the cabling from the insulation standpoint.

With electronic applications with precision timing the velocity of signal propagation is a very important design factor. In these instances, waveform distortion can result if the propagation velocities are low. Thus, the difference between accurate data transmission and high amplitude noise has often been traced to waveform distortion.

## 7.2  WIRE AND CABLE MATERIALS [2.3]

The selection of the appropriate wire and cable begins with consideration of the types of conductors, insulation, shielding and jacketing to be used. Their design parameters and materials will influence the end product's suitability for a particular application. Once these have been properly chosen, the size, shape, form factor, etc., of the interconnection wiring assembly can be taken into account.

### 7.2.1  Wire Conductors

The selection of the appropriate conductor depends on several parameters. Included among these are its base material, coating, construction, stranding, and current carrying capacity. Cost criteria must also be taken into account.

#### 7.2.1.1  Wire Conductor Base Material

Copper is by far the most widely used conductor base material. Among its physical properties are high levels of electrical and thermal conductivity, ductility, malleability, solderability, melting point, and resistance to corrosion, wear and fatigue, Table 7.3. Basically there are two copper conductor materials, i.e., electrolytic tough-pitch (ETP) copper and oxygen-free high-conductivity (OFHC) copper. Except for solderless (wire) wrap applications where improved ductility is significant, the majority of electronic applications use the ETP type.

Copper covered steel combines the conductivity and corrosion resistance of copper with the strength of steel. The actual conductivity of this type of conductor depends on circuit frequency. At high frequencies it is the same as copper due to the skin effect. At lower frequencies it is significantly lower.

High-strength alloys, although they are more expensive than the copper and copper covered base materials, are used in applications that require significant size and weight reductions. They also offer high breaking strength and greater "flex" life with only a small increase in direct current resistance.

The substitution of aluminum for copper has received some consideration for potential weight and cost reduction purposes. Aluminum has many properties that are similar to those of copper. However, its use in small gauge wire has been limited. The reasons for this include:

- Restricted miniaturization, because its size must be increased by more than 50 percent for the same current-carrying capacity
- Soldering difficulty, with a danger of galvanic corrosion when it is soldered to another base material
- Cleaning difficulties, associated with the removal of oxide films prior to termination
- Minimal cost reduction when the increase in equivalent wire size is taken into account.

**TABLE 7.3.** Typical Properties of Conductor Materials [3]

| Conductor material | Minimum conductivity (%) | Tensile strength (p.s.i.) | Tensile strength (kg/cm²) | Elongation* (%) | Operating temperature (°C) | Resistance to Oxidation | Resistance to Galvanic corrosion | Resistance to Solderability | Relative weight |
|---|---|---|---|---|---|---|---|---|---|
| Copper | | | | | | | | | |
| Annealed, bare | 100.00 | 35,000 | 2461 | 15–30 | 150 | Poor | Good | Fair | 1.000 |
| Annealed, tinned | 97.16 | 35,000 | 2461 | 10–25 | 150 | Good | Good | Good | 1.000 |
| Annealed, silver coated | 100.00 | 35,000 | 2461 | 15–30 | 200 | Good | Poor | Good | 1.000 |
| Annealed, nickel coated | 96.00 | 35,000 | 2461 | 15–25 | 260 | Good | Good | Poor | 1.000 |
| Medium hard drawn, bare | 96.66 | 55,000 | 3867 | 0.88–1.08 | 150 | Poor | Good | Fair | 1.000 |
| Hard drawn, bare | 96.16 | 65,000 | 4570 | 0.85–1.06 | 150 | Poor | Good | Fair | 1.000 |
| Copper-covered steel | | | | | | | | | |
| Annealed, bare | 40.00 | 47,000 | 3304 | 10–15 | 200 | Good | Good | Fair | 0.925 |
| Annealed, silver coated | 40.00 | 47,000 | 3304 | 10–15 | 200 | Good | Fair | Good | 0.925 |
| Hard drawn, bare | 40.00 | 115,000 | 8085 | 1–1.5 | 200 | Poor | Good | Fair | 0.925 |
| Hard drawn, bare | 30.00 | 135,000 | 9491 | 1–1.5 | 200 | Poor | Good | Fair | 0.925 |
| High strength alloy, silver-coated | | | | | | | | | |
| Cadmium-chromium copper† | 85.00 | 55,000 | 3867 | 6 | 200 | Good | | Good | 0.980 |
| Cadmium copper | 80.00 | 85,000 | 5976 | 0.82–1.06 | 200 | Good | | Good | 0.980 |
| Aluminum, EC grade | | | | | | | | | |
| Bare | 62.00 | 12,000 | 844 | 23 | 150 | Poor | | Good | 0.304 |

*Values are for solid wires 8 AWG and smaller, and for individual strands before stranding. Varies with diameter.
†Alloy 135

### 7.2.1.2   Wire Conductor Coatings

Bare copper is rarely used in electronic applications since it will oxidize when exposed to the atmosphere, and thus, impair its ability to be reliably terminated. Some insulating materials also tend to corrode the bare copper.

For temperatures below 150°C the least expensive protective coating is tin. However, tin should not be used above 150°C because it will oxidize rapidly.

Silver coating is generally used up to a maximum of 200°C for these higher temperature applications and as required for the use of certain high-temperature insulation materials. Thus, silver coatings are widely used with fluorocarbon, polyimide and silicone rubber insulation.

Silver is readily solderable. Unfortunately this can be a disadvantage because solder can flow under the insulation and potentially reduce flex life at the termination end of the conductor. The silver coated copper is also susceptible to corrosion that is caused by the galvanic interaction between the dissimilar metals.

Nickel is a good copper protective coating material for continuous operating temperatures up to 300°C. Another advantage of this material is that it is not susceptible to the corrosion potential of silver. However, its main disadvantage is poor solderability even with the use of activated fluxes and high soldering temperatures.

### 7.2.1.3   Wire Conductor Construction

The construction plays a significant role in the proper functioning and reliability of a conductor. The consideration of construction involves deciding whether the conductor is to be solid or stranded, and if stranded, what type of stranding is to be employed.

A.   Solid Wire.   A solid wire construction is relatively low in cost and weight. Unfortunately, it will not withstand mild flexure without breaking because of working hardening and fatigue. Thus, it is used primarily in electronic solderless (wire) wrap applications.

B.   Stranded Wire.   Various wire-stranding configurations, Figure 7.1, have been developed in order to overcome the disadvantages associated with using solid conductor wire. However, the construction characteristics of stranded wire may limit the applications of the insulation materials.

"Bunch" stranding consists of a twisted group of strands which have the same length. It offers extreme flexibility and relatively low cost. But it does not provide a consistent circular cross-section and is independent of geometric arrangement. Therefore, the strands are susceptible to movement, a condition which is not recommended for use with an extruded, thin-wall, insulated-wire construction.

The "true concentric" construction has the most consistent circular cross-

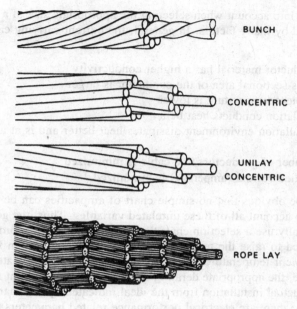

**Figure 7.1.**    Types of conductor stranding. [1]

section of all of the available strandings. This is achieved by having alternate layers of conductors applied in opposite directions with increasing lay. This holds the strands in place and prevents strand "popping" and the creation of "high" strands. Thus, the concentric construction is preferred for use with extruded, thin-wall, insulated wire.

The "unidirectional concentric" construction differs from the true concentric arrangement in that the lay of successive layers is applied over a core, in the same direction. This gives the wire more flexibility and endurance.

A further improvement in this regard can be achieved with the "unilay" construction along with a relatively smaller wire diameter. "Equilay" is a variation of the concentric stranding construction that has a reversed, but equal length, lay and a stiffer construction than does the unidirectional arrangement.

The "rope" lay is basically a large-gauge, no.10 AWG and larger, conductor construction that consists of a central core member. The other strands are formed about the center conductor in a bunch or concentric manner. This results in a stranded wire construction with a uniform circular cross-section and good flexibility.

### 7.2.1.4   Wire Conductor Size

Ampacity or current carrying capacity, the current a conductor can carry before its temperature rise exceeds a permissible value, is one of the initial factors that must be taken into account when selecting the appropriate size of a conductor. It is influenced by many factors. In general, the current-carrying capacity will increase when:

- The conductor material has a higher conductivity
- The cross-sectional area of the conductor is larger
- The ambient temperature is lower
- The insulation conducts heat better
- The installation environment dissipates heat better and is at a lower altitude
- The number of conductors in a cable is minimized
- The applied current (amperage) is minimized.

It should be obvious that no simple chart of ampacities can be constructed that takes into account all of these unrelated variables. Thus, the general practice is to initially use a selection chart that merely shows the approximate amperage required to raise the temperature of a single conductor in free air at a specified ambient temperature for commonly used insulating materials. With this as a guide, the appropriate derating should be used to account for the variations of the actual installation from the ideal indicated by the chart (or table).

In addition, there are electrical performance related parameters that will affect the selection process. Voltage drop for the anticipated load, continuous duty rating, and short-time rating are included among these. With such considerations it is apparent that the assignment of a current carrying capacity rating to a conductor is a rather inexact procedure. In addition to all of the factors just mentioned, the packaging engineer may have other concerns, such as service life. Thus, the ampacity may be further derated to provide for an even greater margin of safety.

### 7.2.1.5  Wire Cost Considerations
In general, the finer the size of the individual wire strands in a conductor, the higher the cost. Coated or plated wires are also more expensive. Also, the shorter the length of lay (twist) of a stranded conductor, the greater the flexibility and the higher the cost.

Some stranded conductors can be compacted (passed through a swaging die) to form a more rounded contour. This too increases cost and reduces flexibility.

### 7.2.2  Wire and Cable Insulating Materials [1]

Basically there are two types of wire and cable primary insulations available for electronic applications, i.e., elastomers and plastics. The elastomers used are electrical-grade thermosetting rubbers and rubber-like materials. The plastics are represented by a wide range of thermoplastic resins.

Each of this wide variety of insulating materials has its own advantages and disadvantages when compared to the others. Thus, the selection process can become quite involved. However, there are usually a few basic advantages that make one material the best choice for a given application.

The factors governing the selection of an optimum insulation material cover a wide range and are less theoretical in nature than those for choosing a conductor. In addition to meeting specific system electrical and environmental criteria, related fabrication techniques and assembly must be considered. The insulation often must withstand mechanical abuse and the heat from soldering or related operations. Furthermore, when applicable, the insulation must be compatible with encapsulants, coatings, adhesives, and marking inks.

The properties of insulating (and jacketing) materials used with wiring and cabling in electronic applications can be grouped into mechanical/physical, electrical, and chemical categories. The mechanical and physical properties often considered are:

- Tensile strength
- Elongation
- Specific gravity
- Abrasion resistance
- Cut-through resistance
- Low-temperature (cold bend) stability
- High-temperature (deformation) stability.

The electrical properties include:

- Dielectric strength
- Dielectric constant
- Loss factor
- Insulation resistance.

Among the chemical properties are:

- Fluid resistance
- Flammability
- Temperature resistance
- Radiation resistance.

### 7.2.2.1  Rubber Insulation

The electrical-grade rubber elastomers are elastic materials that, because they are thermosetting resins, will not appreciably soften, drip, flow, or deform when exposed to heat. As compared to the thermoplastic insulating resins, the rubbers also have less inherent stiffness, so that they are more flexible and have the ability to lie flatter, and are generally less likely to stiffen or crack when exposed to cold temperatures.

However, for electronic applications the thermoplastic insulations tend to have better electrical properties, lower costs, greater choice of color, lighter weight, and can be applied with thinner walls. The rubber type of insulation is usually more difficult to strip cleanly because of its adhesive qualities.

Thus, a separator of paper tape or textile serving must be applied between the conductor and the rubber insulation to keep the insulation from sticking to the conductor during the extrusion process. This adds to manufacturing costs and still does not make the rubber insulation as easy to strip cleanly as is the thermoplastic insulation.

**A.   Low-Cost Rubbers.**   Synthetic natural rubber, styrene-butadiene rubber (SBR), and polybutadiene rubber, Table 7.4, are the lowest in cost and are commonly are used alone or in blends with one another for use as the primary wire insulation or the overall protective jacket. However, their ozone resistance is poor unless anti-ozone additives are used. Unfortunately, these additives usually cause straining.

**B.   Butyl, EPR and EPT Rubbers.**   The use of butyl rubber is often recommended for high-voltage applications because of its freedom from porosity and excellent ozone and corona resistance. Ethylene-propylene copolymer (EPR) and ethylene-propylene terpolymer (EPT) rubbers also have excellent ozone and corona resistance.

**C.   Silicone Rubber.**   Silicone rubber compounds are ideal for severe conditions in which resistance to temperatures up to 260°C. flexibility down to −75°C. excellent corona resistance, and outstanding resistance to weathering, ozone and chemicals are needed. However, silicone rubber costs more than the other rubber insulations. Also, some formulations are not very strong mechanically and, thus, have a low cut-through resistance.

**D.   Other Rubbers.**   The polychloroprene (Neoprene) and chlorosulphonated polyethylene (Hypalon®) type of rubber are quite similar and have excellent oil, flame, heat, ozone and weather resistance. Nitrile rubber (NBR) is suggested for use in applications that require resistance to oil, gasoline, or heat. However, its generally inferior electrical properties often limit its use in electronic products.

### 7.2.2.2   *Plastic Insulation*
The variety of plastic wire and cable insulating materials is as numerous as the rubber types, Tables 6.5 and 7.5. However four basic thermoplastic polymers serve the bulk of the needs for electronic applications, i.e., polyvinyl chloride, polyethylene, polypropylene, and the fluoropolymer resins.

| Property | Styrene-butadiene (SBR) | Natural | Synthetic natural | Polybutadiene | Neoprene | Chlorosulfonated polyethylene (Hypalon®) | Nitrile (NBR) or butadiene-acrylonitrile | Ethylene-propylene copolymer (EPR) | Ethylene-propylene terpolymer (EPT) | Butyl | Silicone |
|---|---|---|---|---|---|---|---|---|---|---|---|
| Oxidation resistance | F | F | G | G | G | E | E | E | E | E | E |
| Heat resistance | F-G | F | F | F | G | E | G | E | E | G | O |
| Oil resistance | P | P | P | P | G | G | G-E | F | F | P | F-P |
| Low temperature flexibility | F-G | G | E | E | F-G | F | F | G-E | G-E | G | O |
| Weather, sun resistance | F | F | F | F | G | E | F-G | E | E | E | O |
| Ozone resistance | P | P | P | P | G | E | G | E | E | E | O |
| Abrasion resistance | G-E | E | E | E | G-E | G | G-E | G | G | F-G | F |
| Electrical properties | E | E | E | E | F | G | P | E | E | E | O |
| Flame resistance | P | P | P | P | G | G | F | P | P | P | F-G |
| Nuclear radiation resistance | F | F | F | P | F | F-G | F-G | F | F | P | F |
| Water resistance | G-E | G-E | E | E | G | G-E | G-E | G-E | G-E | G-E | G-E |
| Acid resistance | F-G | F-G | F-G | F-G | G | E | G | G-E | G-E | E | F-G |
| Alkali resistance | F-G | F-G | F-G | F-G | G | E | F-G | G-E | G-E | E | F-G |
| Gasoline, kerosene, etc. (aliphatic hydrocarbons) resistance | P | P | P | P | G | F | F | P | P | P | P-F |
| Benzol, toluol, etc. (aromatic hydrocarbons) resistance | P | P | P | P | P-F | F | P-F | F | F | F | P |
| Degreaser solvents (halogenated hydrocarbons) resistance | P | P | P | P | P | P-F | P | P | P | P | P |
| Alcohol resistance | F | G | G | F-G | F | G | E | P | P | E | G |

P = POOR   F = FAIR   G = GOOD   E = EXCELLENT   O = OUTSTANDING

These ratings are based on average performance of general purpose compounds. Any given property can usually be improved by the use of selective compounding.

155

TABLE 7.5. Typical Properties of Thermoplastic Insulation Materials [1]

| | Polyvinyl chloride (PVC) | Low-density polyethylene | Cellular polyethylene | High-density polyethylene | Polypropylene (PE) | Polyurethane (PU) | Nylon | Fluoro-polymers (Teflon®) |
|---|---|---|---|---|---|---|---|---|
| Oxidation resistance | E | E | E | E | E | E | E | O |
| Heat resistance | G-E | G | G-E | E | E | G | E | O |
| Oil resistance | E | G-E | G-E | G-E | E | E | E | O |
| Low temperature flexibility | P-G | G-E | E | E | P | G | G | O |
| Weather, sun resistance | G-E | E | E | E | E | F-G | E | O |
| Ozone resistance | E | E | E | E | E | E | E | E |
| Abrasion resistance | F-G | F-G | E | E | F-G | O | E | G-E |
| Electrical properties | F-G | E | E | E | E | P-F | F | E |
| Flame resistance | E | P | P | P | P | P | P | O |
| Nuclear radiation resistance | P-F | G | G | G | F | G | F-G | P-F |
| Water resistance | E | E | E | E | E | P | P-F | E |
| Acid resistance | G-E | G-E | G-E | G-E | E | F | P-F | E |
| Alkali resistance | G-E | G-E | G-E | G-E | E | F | E | E |
| Gasoline, kerosene, etc. (aliphatic hydrocarbons) resistance | G-E | P-F | P-F | P-F | P-F | F | G | E |
| Benzol, toluol, etc. (aromatic hydrocarbons) resistance | P-F | P | P | P | P-F | P | G | E |
| Degreaser solvents (halogenated hydrocarbons) resistance | P-F | P | P | P | P | P | G | E |
| Alcohol resistance | G-E | E | E | E | E | P | P | E |

P = POOR   F = FAIR   G = GOOD   E = EXCELLENT   O = OUTSTANDING

These ratings are based on average performance of general purpose compounds. Any given property can usually be improved by the use of selective compounding.

156

A.   Polyvinyl Chloride.   Polyvinyl chloride (PVC) compounds are very popular for use in audio-frequency electronic applications because of their high dielectric and mechanical strengths, softness, and resistance to flame. (The latter is especially important in order to satisfy Underwriters' Laboratory (UL) and military flammability requirements.) Also, at high temperatures the PVC insulations will generally cling and maintain their form, thus prolonging circuit continuity. However, the big disadvantage of using polyvinyl chloride in electronic applications is its high capacitive and loss properties. PVC must also be compounded with plasticizers that have been known to migrate into other insulations, especially polyethylene, and thus degrade the electrical properties. This loss (migration) of plasticizer can cause PVC to embrittle and crack. A further drawback is that the polyvinyl chlorides are very sensitive to low-temperatures that can cause them to stiffen appreciably and crack when bent.

The PVC compounds can be formulated to meet special requirements for a broad range of desirable characteristics, but with associated drawbacks. For example, higher temperature (105°C.) vinyls are available. However, most higher-temperature PVC compounds are stiffer and harder than the lower-temperature grades, and thus might be less desirable for a specific application.

B.   Polyethylene.   Polyethylene (PE) is the insulation that is often recommended for most radio-frequency (RF) electronic wiring and cabling applications that require superior electrical characteristics. Because of its dielectric constant, loss factor, and insulation resistance, PE can be matched electrically by some of the fluorocarbons. Its solvent resistance, moisture resistance, and low-temperature performance are superior. For example, unlike polyvinyl chloride, PE-insulated wire can be bent without cracking, even though it is stiff at low temperatures. A disadvantage is its poor flammability rating. Although, the use of additives can alleviate this problem to some degree but with a degradation in electrical properties.

C.   Polypropylene.   A thermoplastic insulation that is gaining in popularity for electronic applications is polypropylene. This is primarily because its electrical properties are somewhat similar to those of polypropylene, and yet it is tougher, lighter and more abrasion and heat resistant. Polypropylene foam insulation has been used in miniature cabling applications because of its toughness and low capacitance.

The lack of low-temperature flexibility is the main disadvantage. Thus, polypropylene should not be used in flexing situations at temperatures below −20°C. or in fixed applications below −40°C.

D.   Fluoropolymers.   Fluorinated thermoplastics or fluoropolymers, such as Teflon®, are outstanding in nearly every performance criteria for electronic wiring and cabling. However, the selection process is slightly complicated by the variety of specialized fluoropolymers that are available for these applications.

- Polytetrafluoroethylene (TFE Teflon)) is undoubtedly the best known and most widely used polymer in the fluorocarbon family. It has excellent electrical, chemical, and thermal properties. Polytetrafluoroethylene maintains mechanical stability at temperatures up to 260°C. and thus can withstand contact with a hot soldering iron for short periods of time without damage. Unfortunately, TFE is not a particularly tough material as its abrasion and cut-through resistance are not highly rated. Other potential disadvantages include its poor corona resistance for high-voltage applications and its mechanical degradation in high-radiation environments. However, the latter improves in oxygen-free atmospheres.

- Fluorinated ethylene propylene (FEP Teflon) polymers are a melt-extrudable counterpart to TFE that has almost identical electrical and chemical properties. However, since FEP has a maximum operating temperature rating of 200°C., its soldering and overall temperature resistance is inferior. When used as a primary insulation fluorinated ethylene propylene polymers offer advantages with respect to low cost (in production quantities), longer continuous lengths, and a better ability to be marked using hot stamping.

- Ethylene tetrafluoroethylene (ETFE Tefzel®) is 75% TFE by weight, and thus is also well suited for electronic applications. It can withstand an unusual amount of physical abuse during and after installation, has very good electrical characteristics, such as a low dielectric constant of 2.6 that is nearly constant with changing frequency and temperature, and a dissipation factor that is considerably lower than that of any other extrudable wire insulation material except the Teflons. ETFE's strength compares to that of polyvinylidene fluoride (PVDF) and it is not quite as stiff. Its high flexure life, exceptional impact strength, and service temperature of 150°C. are better than those of PVDF. Other advantages of the material include its good flame resistance, good high- and low-temperature properties, and chemical inertness. For example, it has no known solvent below 200°C. Polyvinylidene fluoride (PVDF Kynar®) is extremely tough and has excellent abrasion and cut-through resistance. As previously mentioned, its electrical, thermal, and chemical properties are inferior to those of TFE and FEP. However, PVDF has a superior radiation resistance. In addition to being used by itself as an insulation, polyvinylidene fluoride has also been combined with polyethylene (polyalkene) resins and used as a jacketing material.

### 7.2.3   Wire and Cable Shielding [2, 4]

Simply defined, a shield is a conductive envelope that is placed around an insulated conductor, or group of conductors. Its purpose is to prevent the external fields from adversely affecting the signals transmitted over the center conductor, to prevent undesirable radiation of a signal into nearby or adjacent conductors, or to act as a second conductor in matched or tuned lines.

Under normal operating conditions, in shielded cables substantially every point on the surface of the enclosed insulated conductor is at ground potential or at some predetermined potential with respect to ground.

A shield's effectiveness is measured by the degree to which electrostatic or electromagnetic energy is attenuated when passing through it. The effectiveness is the summation of the shield's reflection, absorption, and multiple boundary losses. It may be mathematically stated as the ratio of transmitted to incident field energy. Expressed in decibels, and varying with frequency, this ratio can be established for either electrostatic fields, electromagnetic fields, or the composite fields.

#### 7.2.3.1   Wire and Cable Shield Design

In designing a wiring or cabling shield system, such details as line termination, impedance, field strength, and frequency bands of interference are seldom available. Even if approximate values are known, they may prove to be substantially different from what was anticipated at the cable installation.

As a practical matter, therefore, shield design becomes increasingly dependent on empirical data; quantitative effectiveness standards that specify shield reflection and absorption levels are rarely used. Thus, cabling shield design entails identifying the potential fields in which the circuit will operate and then choosing the combination of materials and wiring configuration that will provide an effective barrier against anticipated frequency bands of interference. However, this cannot be done in isolation, for any chosen approach must also be consistent with other cabling requirements, such as flexibility, installation conditions, and cost. Unfortunately, the shielding approach required to minimize the effects of electrostatic fields differ from those found to be effective against electromagnetic field interference.

#### 7.2.3.2   Electrostatic Shielding

Electrostatic interference is relatively easy to guard against. The basic types of electrostatic wiring shields are based on the use of either tapes, wires, conductive plastics or yarns and metal tubes, Figure 7.2. Each of these has a preferred use in some specific application. Thus, the selection of any one type is influ-

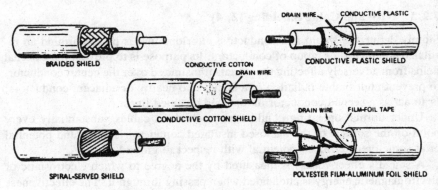

**Figure 7.2.**    Types of wire and cable shielding. [1]

enced by considerations such as shield integrity, flexibility and current capacity, Table 7.6. Their effectiveness in doing this is determined by:

- The degree of shield coverage, 100% is desirable
- The nature of the shielding material, a reasonably good conductor is essential
- Grounding, grounded shields reduce interference by about three-fourths, ungrounded shields by about one-half.

A.  Flat Tape Shields.  Flat all-copper or aluminum tapes, applied either spirally or longitudinally, are often used in electronic applications. However, as a general rule, a combination of a thin aluminum foil laminated with polyester is

**TABLE 7.6.    Typical Properties of Cable Shielding Techniques [1]**

|  | Shield type | | | | |
| --- | --- | --- | --- | --- | --- |
|  | Copper braid | Copper wrapped | Conductive textile | Aluminum mylar | Conductive plastic |
| Shield effectiveness at audio frequencies | Good | Good | Fair | Excellent | Good |
| Shield effectiveness at radio frequencies | Good | Poor | Poor | Excellent | Poor |
| Normal percent of coverage | 60–95% | 90–97% | 100% | 100% | 100% |
| Fatigue life | Good | Fair | Excellent | Good | Good |
| Tensile strength | Excellent | Fair | Good | Good* | Poor |
| Termination method | Comb and pigtail | Pigtail | Drain Wire | Drain Wire | Drain Wire |

*Includes drain wire.

preferred because its use minimizes the outside diameter of the cable. Also, the metal/polyester combination has greater tensile strength than a thin metal foil alone. This provides for additional mechanical stability and reduces the occurrence of dielectric breakdown between adjacent conductors.

Tape shields are easily terminated by the inclusion of a drain wire that is placed in contact with, and along the entire length of, the shielding metal.

**B.  Corrugated Tape Shields.**   Since the amount of metal in flat aluminum/ polyester tape shields is limited, the use of corrugated metal shields is recommended where electrostatic field conditions indicate the need for a larger mass of metal. Thus, corrugated shields of copper, aluminum and copper/stainless steel/copper combinations are used occasionally in electronic cables.

The use of corrugation increases the tape's strength and flexibility. It also provides improved electrical performance by:

- Permitting low-resistance, high-strength termination, because the entire shield cross-section is present at the cable ends
- Increasing the mean distance between the shield and the contained conductors, resulting in reduced interference from induced shield currents
- More closely coupling the fields resulting from the currents and concentrating them at a greater mean distance from the contained conductors.

However, on the negative side, the use of corrugated shields can add approximately 1.8 mm (0.070 inch) to the outside diameter of the cable.

**C.  Braided Wire Shields.**   Shields in the form of braided small-size wires are most effective at frequencies between 1 and 140 kHz. Because they are woven, braided shields retain their structural integrity even without a jacket.

The shielding effectiveness of a braided-wire shield is generally proportional to the amount of coverage. From an electrical standpoint 100% coverage is unattainable, since the braid has minute openings where the strands cross where leakage can occur. However, for the majority of audio-frequency applications a 75 to 85% coverage will generally prove to be effective. At higher frequencies a coverage of from 85 to 95% will probably be necessary.

**D.  Spiral Wire Shields.**   Spiral shields are composed of wires that are served, i.e., applied in one direction by using only one ground of braided carriers. With this construction there is no interweaving of the strands; all shield wires are parallel and lie adjacent to one another.

The primary advantages of using spiral shields are that they are more flexible than braided shields, and they minimize the amount of manual labor needed to terminate them. When the jacket is removed, the strands, being loose, need only to be twisted together and terminated.

Although greater than 95% coverage is obtainable with spiral shields, they are not as effective as braided shields at higher circuit frequencies. Thus, they are recommended only for audio-frequency applications because of the "coil effect" produced by the inductance of the served wire strands.

E.  Reverse-Spiral Wire Shields.  Reverse-spiral shields are formed by placing one or more copper strands underneath a spiral shield. There is no interweaving of wires in this construction, merely two individual layers applied in opposite directions. The use of corrugated copper, aluminum or bimetallic tape are the most effective. Next in decreasing order of effectiveness are spiral copper, aluminum/polyester tape, braided copper, conductive textile, and conductive plastic.

F.  Conductive-Plastic Shields.  Polyvinyl chloride (PVC) and polyester compounds have been formulated with conductive adhesives for the purposes of shielding. Unfortunately, their effectiveness has been relatively poor, and thus normally limited to the low audio-frequency range. As with foil, termination is also a problem; however, a drain wire can be used.

G.  Conductive-Yarn Shields.  Electrically-conductive yarns, such as impregnated glass, provide weight reduction. Unfortunately, a lack of good conductivity and difficulty in termination makes them suited only for special applications.

H.  Metal and Conductive-Yarn Shields.  A combination of metal strands and conductive yarns has been developed which when braided, provides effective shielding and reduced shield weight. Termination of this type of shield can be accomplished by normal techniques.

I.  Solid Shields.  Solid or tubular shields have been applied by tube swaging, interlocked forming, seamwelding, or the use of soldered tape around the dielectric. These approaches offer 100% shielding and can also serve as a protective armor. The solid shields are usually applied to buried cables or where a total shield is mandatory.

### 7.2.3.3  Wire and Cable Electromagnetic Shielding
Shielding against electromagnetic fields presents altogether different problems and solutions. Unfortunately, the materials that are effective in attenuating electromagnetic fields (high-permeability and saturation-flux-density metals) are neither physically nor economically suitable for wire shielding purposes. Thus,

among the more available and suitable materials are heavy layers of steel tape or special magnetic alloys, such as Permalloy®, and thick corrugated bimetallic and copper tapes.

The easiest and generally the best way to protect flexible cabling against magnetic disturbances is to twist the circuit's conductors closely together in order to electrically cancel the effect of the field. The balanced twisted pair is ideal in this regard. Where strong electromagnetic fields are present, thick metal shields and tightly twisted pairs of wires should be used together.

Another method for minimizing electromagnetic problems is to shield the cable with a ferromagnetic material that conducts the field around the cable, rather than through it. However, such a cable is relatively costly, thick, and inflexible. Furthermore, unless the shield is relatively thick, it can saturate quickly, and thereby leak some of the field through the cable. In a fixed installation, the distribution of cabling in a soft iron pipe offers considerable resistance to electromagnetic interference. [1]

**A.  Direct-Current and Low-Frequency Shields.**  Magnetic materials are effective against direct-current and low-frequency magnetic fields, since they tend to shorten the flux lines that attempt to extend themselves through the shield. Thus, electromagnetic shields of high permeability metals should be used at audio or power frequencies (up to 60Hz) where conductive shields would, otherwise, have to be excessively thick to be effective.

**B.  Radio-Frequency Shields.**  At radio frequencies the most practical electromagnetic shields are those made of materials having low electrical resistivity (high conductivity), such as copper or aluminum. In attempting to pass through such shields the magnetic flux induces voltages in the metal, giving rise to eddy currents. These currents oppose the action of the magnetic flux, thus preventing its penetration through the shield.

In some critical low-level applications it is sometimes necessary to employ alternate layers of high-permeability and high-conductivity shields. In this case, physical separation of the shield layers by dielectric materials is important, particularly when interference from high-current power sources is present.

**C.  High-Frequency Shields.**  At higher frequencies, i.e., up to 1 MHz, the use of copper and steel tapes is the most effective. Next in decreasing order of effectiveness are aluminum tapes and copper braids.

### 7.2.4  Cable Jacketing [1]

Most electronic cables have an overall plastic or rubber jacket that protects the conductors and holds them together. Although jacketing materials are often the same as those insulating the individual conductors within the cable, Table 7.5, they are not compounded for their electrical properties, but for their resistance to physical damage and the effects of the end product environment.

### 7.2.4.1  Rubber and Plastic Cable Jackets

Although seldom done, putting a plastic jacket over rubber-insulated conductors poses few problems, the reverse is not true. This is because a rubber jacket must be vulcanized at high temperatures immediately after it is extruded. However, special techniques are sometimes used to perform this operation over low-temperature insulated conductors.

The degree of tightness of rubber jackets is not as variable as it is with plastic jackets. With plastics, the tightness can be varied to provide a balance among flexibility, flexure life, strippability, and toughness.

### 7.2.4.2  Full Extruded and Tubed Cable Jackets

Most jackets are either permanently full extruded onto the pre-insulated wires in the cable or prepared in the form of tubing that encloses all of the wires within it, Figure 7.3. An extruded jacket is applied under pressure and is forced within and around the conductors. It provides a relatively stiff, abuse resistant, and heavy cable that has an esthetically pleasing, smooth, round outside surface.

A tubed jacket is applied under minimum pressure, has a uniform wall thickness, and is easier to strip. A cable with a tubed jacket is lower in cost, lighter in weight, and much more flexible than a full-extruded jacket. Most plastic cable jackets are tubed. Unfortunately, compared to an extruded jacket, a tubed jacket is not as tough, nor is it as esthetically pleasing because it tends to conform to the unevenness of the cable components within it.

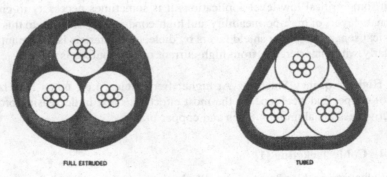

FULL EXTRUDED                          TUBED

**Figure 7.3.**    Full-extruded and tubed jacket configurations. [1]

### 7.2.4.3   Cable Jacketing Migration and Contamination

When selecting a jacketing material it is important to ascertain whether or not it must be nonmigrating (non-marking) to other materials with which it may come into contact. Some of the jacketing materials will mark other surfaces when rested against them for a length of time. For example, one plastic that is very sensitive to marking, especially from vinyl, is polystyrene.

Jackets for radio-frequency (RF) cables are sometimes required to be non-contaminating. This is because plasticizers for some vinyl jacketing compounds will flow into the polyethylene dielectric underneath, and thus cause increased electrical losses.

Unfortunately, the terms nonmigrating and noncontaminating are often confused. To avoid misunderstandings, it is important that the packaging engineer clearly specify the application and end product requirements when involved in the jacketing selection process.

### 7.2.4.4   Other Cable Jacketing Considerations

It is important that the cable jacket be compatible with the molding or encapsulating processes that may be associated with the final cable assembly and termination operations. For example, many jacket compounds will not form a chemical bond with some potting compounds.

It is also important to match the operating temperatures and other environmental factors of the conductor insulation and cable jacketing materials. Avoid the pitfalls that are associated with assuming that they are compatible.

Cables to be installed in underground applications may have to be specified as being rodent-proof depending on where they are used. Some chemical additives can be used that furnish temporary repellency, but none will last for many years. Thus, it is often recommended that a steel armor tape be applied over the cable jacketing.

## 7.3   WIRE AND CABLE CONFIGURATIONS [1, 3]

There are many different configurations of wire and cable used in electronic applications. They usually fall into one of the following categories:

- Round single-conductor insulated wire
- Twisted pairs of round single-conductor insulated wire
- Round coaxial cable
- Round jacketed multiconductor cable
- Flat round-conductor cable
- Flat flat-conductor cable
- Fiber-optic cable.

There are also composite cables that include several of these configurations under one jacket or harnessed assemblies into various custom arrangements.

Variations of these basic constructions are numerous and the terminology used to describe them can be confusing. To add to the confusion they are sometimes referred to by their intended application and not by their basic configuration. Thus, some cables, usually multiconductor round cables, can be referred to as "instrumentation cables" in one instance, and as "communications cables" in another. As will be clarified later, there are also flat "ribbon" cables and flat tape cables.

## 7.3.1  Wire and Cable Standards

Electronic wires and cables are manufactured to either industry standards, government standards, user specifications, or occasionally, to the manufacturer's own internal requirements. In addition, there are standards agencies, such as the American Society for Testing Materials (ASTM), that are concerned mainly with the evaluation of the materials used in the interconnection wiring. Other concerned organizations include:

- The Institute of Electrical and Electronic Engineers (IEEE), which publishes testing and use standards
- The American Society of Automotive Engineers (SAE), which assists United States government agencies in the preparation of standards
- The Rural Electrification Administration (REA), which develops telecommunications standards
- The Insulated Power Cable Engineers Association (IPCEA), which issues detailed construction standards for power and control cables
- The Institute for Interconnecting and Packaging Electronic Circuits (IPC), which develops various discrete wiring usage guidelines and flat cable documents.

There are also other agencies that exercise greater influence on electronic wire and cable usage for the end product equipment and the interconnections between them. Included among these are:

- The Underwriters Laboratories (UL) Inc. and the Canadian Standards Association (CSA) for commercial and industrial products
- Various government agencies throughout the world, including the United States Department of Defense with its "Mil Specs."

## 7.3.2  Round Single-Conductor Insulated Wire

Discrete round single-conductor insulated wire, sometimes called hook-up wire is used to provide point-to-point electrical interconnections between a wide variety of locations within electronic equipment. It is often used:

- In mechanically protected areas, such as in stranded chassis wiring and solid-conductor backplane wiring
- As a basic component of multiconductor cables or wiring harnesses.

Many different types of discrete single-conductor insulated wire are manufactured so that they are in accordance with the requirements of MIL-W-76, MIL-W-16876, MIL-W-81044, MIL-W-81381, etc. Specifications such as these allow, or make provisions for, a variety of primary insulation, conductor platings, strandings, shielding, and jacketing that are suitable for protection against several different mechanical, thermal, and chemical environments.

### 7.3.3  Twisted Pairs of Wires

In general, twisted pairs of individual single-conductor insulated wires are the least expensive type of signal-transmission cable. They come in a variety of insulations and sizes, and although often unshielded and unjacketed, offer a choice of physical and electrical properties at a range of price levels.

However, because of their spiralling configuration, they are difficult to terminate, manual methods are slow, fully automatic procedures are limited to special applications. Despite this drawback, their low price makes them attractive for use in many electronic applications.

Electrically twisted pairs can be grouped into three categories.

- Semirigid wires with a polyvinyl chloride (PVC) insulation
- Unshielded wires with low dielectric constant insulations
- Shielded styles.

The PVC-insulated styles, the ones which are most frequently specified because of their low cost, have broadly toleranced electrical characteristics, such as nonuniform impedance and excessive reflections. There can also be considerable crosstalk at high circuit frequencies when the twisted pairs are stacked together. Thus, this type of wiring is used primarily where electrical properties are not critical.

Twisted pairs employing irradiated-PVC and insulations with a low dielectric constant yield better electrical characteristics at a slightly higher cost. Combinations using low dielectric constant fluoropolymers or cross-linked polyethylene insulation and shielding over individual pairs ("Twinax") meet close electrical tolerance requirements. However, when compared with the use of low cost PVC pairs, these constructions are considerably more expensive. Total installed cost, therefore, becomes especially important when selected twisted-pair cables are used.

### 7.3.4  Coaxial Cable [5]

Coaxial cables can transmit high-frequency cable signals at high speeds with low distortion. They have closely toleranced electrical characteristics, come in a variety of constructions, and have either braided or laminated-metal shields. However, there are alternatives to the use of coaxial cables, such as twisted pairs and fiber-optic cables. Coaxial cable and its connectors are both more expensive than comparable more conventional assemblies. Some of them also have more bulk, less flexibility, and are more difficult to terminate than are other types of signal-transmission cabling. Despite these drawbacks, coaxial cables do serve systems requiring high quality and high speed signal distribution.

There are three predominant classes of coaxial cables for electronic applications:

- Conventional "RG" types, Figure 7.4
- Air-spaced core cables, Figure 7.5
- Cross-linked cellular-polyethylene-core types.

The conventional RG coaxial cable types have been available for several years, are well characterized, and offer a wide spectrum of sizes and performances. They are, for the most part, insulated with solid, foamed, or semisolid polyethylene, solid or foamed polypropylene, or TFE fluoropolymers.

The air-spaced (and foam dielectric) core coaxial cable configurations substantially reduce cable size without sacrificing electrical characteristics. Because they employ an air/flame-retardent polyethylene combination with a very low effective dielectric constant, their electrical properties are superior to those of the conventional RG cables.

**Figure 7.4.**    Typical conventional coaxial cable construction. [4]

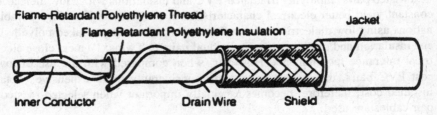

**Figure 7.5.**    Typical air-spaced coaxial cable construction. [4]

Some of the important features of air-spaced coaxial cables are:

- Very low attenuation
- High propagation velocity, a value of 80% as compared to 66% for conventional polyethylene cores and 69% for TFE-core cables
- Easy shield termination, with a silver-coated drain wire that is applied spirally under the shield
- Lower cost, substantial savings as compared to TFE-core cables.

In air-spaced cables, small variations in electrical properties can occur if the air/polyethylene ratio (effective dielectric constant) is changed by mechanical forces, e.g., cable kinking. While this change occurs infrequently, it causes reflections and increases attenuation when it does occur. Though slightly more expensive than air-spaced core cables, the cross-linked cellular polyethylene construction overcomes this problem.

Irradiated cellular-polyethylene coaxial cable is another miniature cable construction with good electrical properties. Its mechanical properties are greatly improved by cross-linking; irradiation increases hardness and resistance to cut-through, soldering iron heat and solvent damage.

### 7.3.5  Round Multiconductor Cable

A round multiconductor cable may be considered as having two or more pre-insulated conductors. Multiconductor cable can have several identical components or a mixture of wire sizes and types contained within a common jacket.

The main features to be considered in the design and manufacture of a typical multiconductor cable, Figure 7.6, include:

- Insulated conductor components
- Cabling or twisting of conductor components
- Shielding
- Fillers
- Binders
- Sheathing
- Armor.

The cable component need not necessarily be an individual insulated wire. It could be a group of insulated wires. It could also be a coaxial or fiber optic cable.

### 7.3.6  Flat Multiconductor Cable [2]

The term flat when used with multiconductor cable refers to the overall configuration of the cable and not to the shape of the conductors within it. Thus, flat multiconductor cables have a basically low-profile rectangular cross-section.

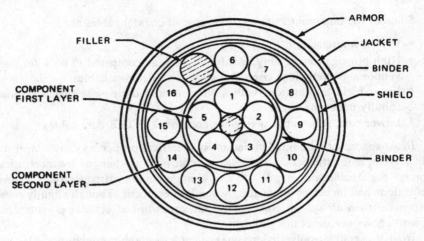

**Figure 7.6.**     Typical round multiconductor cable configuration.

The conductor components of these cables are usually all the same, although mixed combinations can be made. When round conductors are used and the cable is extruded to follow the contour of the round wires, the cable is usually referred to as being a ribbon cable. When flat conductors are used the cable is sometimes referred to as being a tape cable.

All of the various types of flat multiconductor cable have certain characteristics that are inherent in their construction. These features provide performance and application alternatives when compared to the use of round multiconductor cable and when compared to one another, Tables 7.7 and 7.8.

**TABLE 7.7.     Mechanical Performance Features of Cabling Systems [2]**

| Round cabling | Ribbon cable | Flat conductor cable |
|---|---|---|
| Greatest variety of conductor configurations: | Wide variety of conductor configurations: | Limited variety of conductor configurations—some are compromises: |
| Single wires—all gauges | Single wires—limited to smaller gauges | Single conductor—smaller gauges only |
| Shielded wires—all gauges | Shielded wires—limited to smaller gauges | Shielded conductor— smaller gauges only |
| Twisted pair, triples—all gauges | Twisted pair, triples—limited to smaller gauges | Twisted pair—simulated by zigzag crossovers in sandwich |
| Twisted and shielded—all except very large gauges | Twisted and shielded—limited to smaller gauges | Twisted and shielded—simulated in sandwich construction |
| Coaxial cables—all sizes | Coaxial cables—limited to smaller gauges | Coaxial cable not practical— twin-line construction must be used |
| Jacketed overall for heavy duty, if needed | Not suited for jacketed cables | Not suited for jacketed cables |

**TABLE 7.7.**     (*Continued*)

| Round cabling | Ribbon cable | Flat conductor cable |
|---|---|---|
| Round-bundle configuration is relatively self-supporting. | Flat-bundle configuration requires moderate support | Tape configuration requires continuous support |
| Stranded wire usually required for increased mechanical life | Stranded wire is desirable. On larger quantities of conductors per cable, solid wire can be used owing to the mutual mechanical effects of adjacent wires | Foil is ordinarily used. Some modifications have used stranded wire configured to a rectangular cross section |
| Termination preparation—simple mechanical or hotwire stripper | Termination preparation—simple mechanical or hotwire stripper after separating wires from each other | Termination preparation—Minimum practical production process exists for removal of insulation to expose conductors. Certain types of insulation respond to new welding techniques which do not require insulation removal |
| Rugged insulation achieved by selective combination of dielectric materials in layers | Rugged insulation achieved by selective choice of primary dielectric plus an over-all protective ribbon skin. Cables without this skin permit undesired separation of conductors even with careful handling | Edges of tape insulation or complete laminate should be reinforced because, once nicked, the cable can be torn across conductors with relatively small force. A deep scratch or cut in the outer layer of the cable insulation will allow conductors to fracture with sharp bend at the scratch |
| Standard voltage-drop characteristics; use standard current ratings. Bundled circuits require derating because of heat rise if circuits are operated simultaneously | Standard voltage-drop characteristics; use standard current ratings. Derating not normally necessary | Voltage-drop characteristics will vary more widely than standard. Broad, tinsellike conductors permit much higher current limits compared with the equivalent cross section of circular conductors. In general this permits use of much less copper for a comparative current requirement, if the increased voltage drop can be tolerated. |

**TABLE 7.8.    Electrical Performance Feature of Cabling Systems [2]**

| Round cabling | Ribbon cable | Flat conductor cable |
|---|---|---|
| Requires no heat sink in normal application | Requires no heat sink in normal application | Required continuous heat sink in normal application to utilize the high current-carrying capability |
| Interconductor capacitance is high; spacing generally random; capacitance values are unpredictable; capacitance to structure relatively low | Interconductor capacitance is high, spacing fixed, capacitance predictable. Circuits can be selectively spaced to minimize effects. Capacitance from cable to metal support or shielding is high | Interconductor capacitance is low, spacing fixed, capacitance predictable. Circuits can be selectively spaced to minimize effects. Shielding needed in stacked cables. Capacitance from cable to metal support or shielding foil is high |
| Crosstalk is uncontrolled because of random spacing; individual shielding usually required | Crosstalk can be controlled by conductor placement; shielded conductor not usually required. Caution: In stacking cables, overall interlayer shielding may be needed | Crossstalk can be controlled by circuit placement. Overall interlayer shielding may be necessary when cables are stacked |
| Permanently installed interconnections between units and subsystems; also for field cables when jacketed. | Permanently installed interconnections: | Permanently installed interconnections |
| Circular or oval cross-section bundles | Flat-ribbon cross section—ribbons are stackable | Flat tape cross section—tapes are stackable |
| Follows three-dimensional contour without folding or special fabrication, except for very sharp bends | Lies flat against two-dimensional contour, even with sharp bends. Must be folded over or looped for third dimension | Lies flat against two-dimensional contour, even with sharp bends. Foldover preferred over loop for third dimension |
| Requires narrow-width path; can be installed with minimum structural design preparation | Requires moderate width, preferably flat path. Needs detailed structural preparation in initial design | Requires excessive width of continuously flat path. Needs elaborate structural preparation in detail design. During initial design period structure must be designed in detail around cable requirements. |
| Require only air heat sink without special provisions | Requires only air heat sink—higher current ratings achieved with heat sink | Requires continuous heat sink to take advantage of the higher current ratings attainable |

**TABLE 7.8.**    (*Continued*)

| Round cabling | Ribbon cable | Flat conductor cable |
|---|---|---|
| Fixed installation requires ties, lacing, or vinyl jacket | Requires no supplementary bundle ties | Requires no supplementary bundle ties |
| Portable application requires heavy-duty jacket | Not recommended for portable application | Not recommended for portable applications |
| Installation clamps and brackets required; quantity and spacing to suit application | Can be cemented in place for permanent installation or fastened with flat clamps | Should be cemented in place; clamps not recommended unless cable is derated |
| Each conductor requires color coding or stamped identification number. Controlled conductor-to-conductor orientation is not practical | Conductor orientation in cable relative to conductor orientation in an end item should be planned carefully and coordinated during all design phases to achieve proper electrical interface. This is particularly significant when rectangular connectors or grid-spaced terminals are used. Individual conductors can be split apart from the cable as an emergency procedure. Such practice is not recommended | Conductor orientation in cable relative to conductor orientation in and end item must be carefully planned and coordinated during all design phases to achieve proper electrical interface. Individual conductors cannot be reliably split apart from the cable |
| Can be used with largest variety of connectors; i.e., circular, rectangular or special. Termination by crimp, solder, weld pressure, or other techniques | Can be used with large variety of connectors, but the inherent configuration is most compatible with rectangular connectors with terminals spaced on a grid. Grid spacing need not be exactly matched, since the cable conductors may be locally separated one from the other | Variety of connectors is very limited. Grid spacing of connector terminals must match grid spacing of cable conductors. The tolerance control requirements are very exacting. Industry standard is still nebulous |
| Cable replacement is relatively easy; requires loosening or removing clamps and installing new cable without special tools or processes; field repair of individual conductors by simple splice techniques | Cable installed with clamps can be replaced as readily as conventional cables. Use of cable cemented in place is not desirable if field replacement or repair is required. Procedures for removal, cleaning, and reinstallation, and the need for special materials unduly complicate field repairs | Cable installed with clamps can be readily replaced; cable cemented in place cannot be serviced readily. Individual repair is impractical |

The increasing use of flat multiconductor cables in electronic applications arises because of several of their advantages over round configurations, such as:

- Space savings, as much as 70%
- Weight savings, up to 80%
- Flexibility, can be bent, hinged, folded, stretched, or rolled-up
- Reliability, due to consistent electrical characteristics, better heat dissipation, and high abrasion resistance
- Assembly versatility, can be terminated directly to printed wiring board assemblies
- Wiring error minimization, due to their controlled positioning of the conductors with respect to one another
- Ease of identification, conductor components are readily color coded
- Ease of termination, due to their compatibility with mass solderless termination methods
- Cost savings, up to 70% because of the aforementioned features.

### 7.3.6.1  Round-Conductor Flat Cable

Round-conductor flat multiconductor cable is normally constructed using conventional insulated solid or stranded conductors arranged in a side-by-side parallel arrangement. The individual wires are then heat or adhesive bonded together to form a flat ribbon. Variations on this method of fabrication include extruding or calendering, i.e., conductor embedment in an insulating material.

Such cables come in a variety of product types. However, the most common variety of round-conductor flat multiconductor cable is the 10-color repeat type that is composed of groups of ten conductors that are individually color coded and positioned for precise termination and separation. The use of shields and alternating ground conductors result in high performance, transmission-line versions.

Polyvinyl chloride (PVC) is the most common insulation material used in ribbon cable constructions. Such cables have also been made with polytetrafluoroethylene (TFE), and fluorinated ethylene propylene (FEP), polyalkene polyimide film, and polyvinylidene fluoride insulating materials.

Woven cable is a unique form of the ribbon cable construction whereby a weaving process is used to hold together individual insulated conductor components. Since this is based on the use of modifed easily programmable textile weaving equipment, the variations that can be achieved are almost limitless; almost any group of conductor components can be combined in almost any arrangement.

The characteristics of woven cable are essentially identical to those of the basic ribbon cable. However, the weaving technique does provide the following additional advantages.

- Conductor separation—The separating of conductors of a bonded ribbon cable can cause tearing or puncturing of the basic primary insulation. The woven technique permits conductor separation anywhere along the length of the cable without this hazard.
- Folding and stretching—Longitudinal groups of fibers can be incorporated at specific intervals in the woven matrix to provide for natural folding or stretching points. The cable can then be folded in an accordion configuration.
- Mechanical protection—Since the weaving fibers encompass each individual insulated conductor, additional mechanical protection is provided. A strain relief web can also be incorporated into the woven matrix.
- Cabling flexibility—The spacing of the conductors can also be varied, instead of always being in a regular pattern. It is also possible to have a programmed breakout of individual conductors or groups of conductors at branching points along the length of the cable.

### 7.3.6.2  Flat-Conductor Flat Cable

There are two basic types of flat-conductor flat cable, namely, flat-conductor continuous tape or laminated cables, and flexible-dielectric printed circuitry. The major differences between them is their manner of design, construction and application.

A.  Tape Cable.  The tape cable type is constructed by encapsulating flat rectangular conductors between layers of dielectric film. The conductors are usually prepositioned in a regular parallel fashion with uniform spacing, although the use of different conductor widths and spacing is possible within a single cable.

The majority of these types of cables are constructed by roll laminating techniques. However, it is feasible to fabricate continuous lengths of such a cable with printed or etched conductors. Shielding can also be accommodated.

B.  Flexible Printed Circuits.  Flat-conductor flat cables can be made using conventional flexible printed circuit etching processes. Although die-stamping and die-molding processes have been used.

The conventional etching process involves the use of copper foil that has been clad to a flexible thermoplastic dielectric laminate. Traditional printed circuit artwork and subtractive imaging processes are then used to define the customized interconnection wiring pattern. An insulating cover layer can be laminated over the imaged wiring (see Section 5.2.1.8) or insulating coatings may be applied.

The advantages of this approach is, obviously, that almost any interconnection and termination pattern can be produced. However, this is achieved only within the size limits of the fabrication equipment, e.g., not usually in continuous lengths, and with the added setup and tooling costs associated with custom designs.

### 7.3.7    Optical Fiber Cable [6]

In the cable configurations discussed so far the signal transmission medium has been in the form of insulated copper conductors. The use of optical fiber cabling systems offers a radical alternative to traditional signal-transmission systems. However, it is important to keep in mind that the use of fiber optics is basically a system consideration. Thus, instead of just comparing the use of one type of cable or another, the use of fiber optics also involves the design of signal-transmitting and receiving circuitry with optical capabilities and the use of special connectors and terminating techniques. This, perhaps more than any other consideration, has limited the acceptance of fiber-optic technology to a narrow base of electronic applications. With this in mind, there are several advantages that the use of fiber optics offers over conventional systems. Perhaps one of the most important factors to consider in comparing fiber optics with copper-based cabling in the tremendous difference in information capacity or bandwidth between the two approaches. For example, a typical fiber-optic cable has a bandwidth in excess of 500 MHz. This means that an optical cable used in a telecommunications system can handle the entire audio and video range, including cable television, as well as digital information. Another advantage is that with fiber optics the transmitted light signals are not distorted by any form of outside electronic, magnetic, or radio-frequency interference. Also, when compared to the use of coaxial cables with the same signal-carrying capability, the smaller diameter and lighter weight of fiber optic cables allows a relatively easier installation.

In fiber optic systems the fiber type, whether step index or graded, must be decided. The choice is based on a number of criteria, such as the required bandwidth, the length of the transmission line, the operating wavelength, and the power and cost budgets.

An optical fiber is composed of two different optically transparent materials with an outer cover. The center region is the core or light guiding area. Surrounding the core is a region of cladding. The light travels through a fiber by a process known as total internal reflection or TIR. With TIR a ray of light traveling in the core of the fiber reaches the core/cladding boundary and one or two things can happen, Figure 7.7. If the angle of the ray within the boundary is greater than a certain critical angle, which depends on the refractive indices $n$ of the core and cladding materials, the ray will be partially refracted into the cladding and partially reflected into the core. If, on the other hand, the ray is incident on the boundary at an angle less than the critical angle the ray will be totally reflected back into, and connected by, the core.

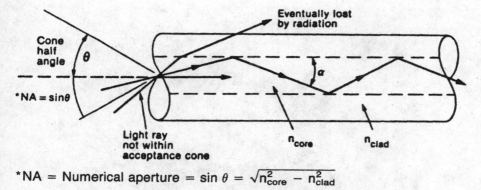

$$*NA = \text{Numerical aperture} = \sin \theta = \sqrt{n_{core}^2 - n_{clad}^2}$$

**Figure 7.7.**     Elements of total internal reflection. [6]

The numerical aperture (NA) value of an optical cable is defined as the maximum angle of incidence of a ray that is totally reflected at the core/cladding interface. As the NA increases more light is transmitted. Thus, in order to facilitate having the most efficient light transmission, the cladding is designed so that it has a refractive index that is lower than that of the core material. This ensures that light traveling through the optical fiber is guided by the core. The outer layer or buffer is a plastic material that provides the fiber with mechanical and environmental protection. The composition of the material used for the core and cladding may vary depending on the specific application. A glass core with a glass cladding is the most commonly used fiber in electronic applications because this combination of materials provides the highest quality of transmission and the lowest optical loss. However, plastic/plastic and glass/plastic arrangements have also been used.

## 7.4  CABLING CONNECTORS [7]

Cabling connectors provide electrical interconnections between a unit and its mounting enclosure (rack) or interface-cabling harness. An alignment device usually is provided to help ensure correct mating.

Coupling devices are not normally involved when both halves of the connector are hard mounted in blind-mating rack-and-panel applications. However, when one, or both, parts of the connector terminate a wiring harness, polarization, keying, and coupling features are generally incorporated.

A wide range of cabling connectors are available to give a good deal of latitude in specifying performance, packaging features, cost, reliability, etc. For example, rectangular D-subminature connectors, Figure 7.8, are engaged by pushing them together, with or without the subsequent use of threaded fas-

**Figure 7.8.**    Push-together coupling rectangular D-submininature discrete wiring connectors. (*Courtesy of AMP Inc.*)

teners or clips to hold them together; circular mulitipin connectors, Figure 7.9, can be engaged and retained together with a threaded or bayonet coupling; circular coaxial-cable connectors, Figure 7.10, can be engaged and retained together with a quarter-turn bayonet coupling; and insulation-displacement (IDC) flat-ribbon cable connectors, Figure 7.11, can be engaged and held together by the use of latches. [8]

Despite the variety of configurations and features available, there are several basic selection considerations that are common to most of them. The most-obvious criteria for selecting cabling connectors are the electrical requirements for the application. In all cases, however, requirements must be broadened to include consideration of production assembly, packaging, field maintenance, and operational environment.

Cabling connectors can provide from a very few to a hundred or more contacts, low- to high-density insert arrangements, a variety of contact sizes and types, and extremes of environmental stability.

The common selection factors can be summarized to include:

- Shell size and type
- Number of contacts

**Figure 7.9.**    Threaded-coupling circular discrete wiring connectors. (*Courtesy of ITT Cannon Corp.*)

- Size of contacts, generally governed by maximum current-carrying requirements
- Contact type, e.g., signal, power, coaxial, fiber-optic, etc.
- Contact material and finish
- Contact construction, i.e., stamped or machined
- Voltage rating, affects insulation material selection and contact spacing
- Contact density (Spacing)

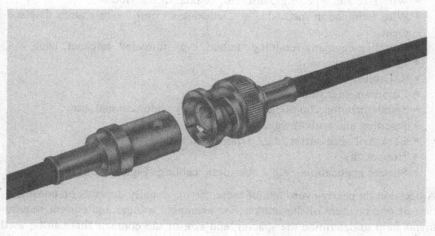

**Figure 7.10.**    Bayonet-coupling, crimp-termination, coaxial cabling connector. (*Courtesy of AMP Inc.*)

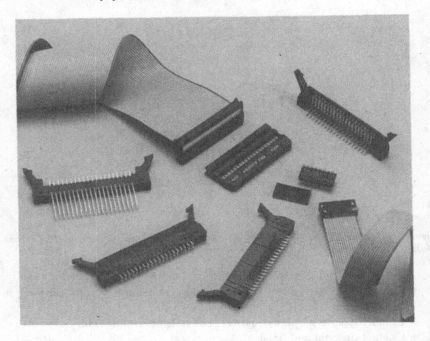

**Figure 7.11.**    Push-together coupling, insulation-displacement termination, flat-ribbon cabling connectors. (*Courtesy of AMP Inc.*)

- Current rating
- Wire size and type, e.g., discrete, coaxial or flat ribbon
- Wire termination method, e.g., solderless crimp or insulation displacement
- Coupling engaging/retaining method, e.g., threaded, bayonet, latch
- Dielectric material
- Reliability
- Environmental stability
- Standardization compliance, e.g., military, international, etc.
- Insertion and withdrawal forces
- Backshell accessories, e.g., strain-relief clamp
- Hermeticity
- Special applications, e.g., fiber optic cabling (Figure 7.12). [9]

A decision on nearly every one of these factors usually depends on consideration of one or more of the others. For example, voltage and current requirements help to determine the spacing and size of the contacts that can be used in a given connector shell size. Also, reliability, environment, ruggedness, etc., influence the choice of materials.

**Figure 7.12.**    Threaded-coupling fiber-optic cabling connector. (*Courtesy of AMP Inc.*)

### 7.4.1  Contact Type

Perhaps the most important consideration in the selection of a specific cabling connector is the type of contact with respect to its wiring termination features. For most conventional low-volume/low-density applications solder-cup contacts are used. However, as usage volume and contact densities increase, there is an increased selection of the solderless contact types. (See Section 6.5).

### 7.4.2  Contact Size

The size of the contact is generally governed by maximum current carrying requirements. Although, cabling connector contacts are available with current ratings suitable for most electronic applications, the current rating of the contact limits the size of the wire to be terminated. For example, a 6-ampere current rating is most often used with a size 22-AWG wire, and a 20-ampere contact with a size 16-AWG wire.

### 7.4.3  Voltage Rating

Voltage rating indicates the highest voltage to which a connector can be subjected without dielectric breakdown or flashover. While connectors are available with ratings up to several thousand volts, most electronic applications require a rating of only 1,000-volts.

The voltage rating will affect several characteristics of the connector. For example, the higher the voltage rating required, the greater the contact spacing must be to prevent electrical breakdown. This will require a larger and heavier connector than might otherwise be possible.

### 7.4.4    Contact Material

The selection of material for cabling connector contacts is based on the necessary performance requirements. Materials with the highest conductivity should be used whenever it is cost-effectively possible in order to keep contact resistance to a minimum. Therefore, beryllium copper, phosphor bronze, and sometimes brass, are used, depending on the properties and costs involved.

The selection of the contact plating is also important. This usually depends upon such conditions as type of circuitry, wear, and environmental requirements.

### 7.4.5    Insulation Material

Two of the many important factors governing the selection of insulation materials for cabling connectors used in electronic applications are operating voltage and altitude requirements. These factors help to determine contact spacing and, thus, creep path distances. Such materials must have good electrical properties as insulators and good mechanical properties as structural members of the connector.

The fabrication of the insulator must be considered in order to optimize performance while minimizing costs. Long-thin designs require the use of high-impact materials. Thin-wall sections, necessitated by close contact spacings, require the selection of a material with good flow characteristics. Resilient materials are usually selected where environmental sealing is required.

### 7.4.6    Backshell Accessories

When selecting a cabling connector it is often important to consider the backshell accessories required. Adapters, cable clamps, covers, sealing rings, etc., must be well matched to the connector design and its characteristics. Other accessories, such as potting shells and sealing plugs, are sometimes needed in special applications.

### 7.4.7    Specifications

When applicable, in the absence of many commercial documents, cabling connectors are usually governed by military specifications. MIL-STD-1353 establishes requirements for the selection of all military connectors and their associated hardware. It is also provides a good start for the selection of connectors for high-performance nonmilitary electronic applications.

For so called rack-and-panel applications where the D-subminiature connector, Figure 7.8, is used the applicable specification in MIL-C-24308. These connectors have found acceptance in several nonmilitary standardization programs, especially those associated with the use of personal computers and other data processing equipment. The circular types, Figure 7.9, are covered by a variety of specifications as listed in Table 7.9.

### 7.4.8  Special Cabling Connector Variations

For special applications there exists a wide selection of special connectors which provide variations of the conventional cabling connectors. For example, for special environments hermetically-sealed connectors can be used; contact variations include the use of filter-contacts and conductive elastomers; and special fiber-optic connectors systems are gaining in popularity for communications applications.

Perhaps the most widely used special connector types are those for use with coaxial cables. These coaxial connectors are used primarily for interconnecting radio- and audio-frequency circuits where stable predetermined values of impedance, capacitance, and/or shielding from external electrical interference must be maintained.

Coaxial connectors are available for cable, Figure 7.10, chassis and bulkhead mounting. Coaxial contacts may also be integrated with certain other cabling connectors.

**TABLE 7.9.    MIL-STD-1353 Circular Connectors [7]**

| Specification number | Title | Contact termination | Contact release | Temp range | Mating technique |
|---|---|---|---|---|---|
| MIL-C-5015 | Connector, electric, MS type (formerly AN) | Solder | fixed/ front/ rear | −55 to 125, 175 200°C | Thread |
| MIL-C-26482 | Connector, electric, circular, miniature, quick disconnect, environment resisting | crimp solder | fixed/ front/ rear | −55 to 125°C | Bayonet Push-pull |
| MIL-C-38999 | Connector, electric, circular, miniature, high density, quick | crimp | rear | −65 to 200°C | Bayonet |
| MIL-C-83723 | Connector, electric, circular environment resisting | crimp and solder | rear/ closed entry | −55 to 175°C −65 to 200°C | Bayonet Thread |

To select the coaxial connector that best matches the requirements of the application and the coaxial cable that is being used, the objective is to provide an uninterruped interconnection path. Therefore, it is desirable to select as simple a connector as possible with electrical properties that closely match those of the cable, or vice versa.

Other important selection criteria include:

- The connector series available for the chosen cable size
- Electrical requirements, such as impedance matching
- Coupling method desired, i.e., bayonet, threaded, or quick-disconnect push-pull
- Wire termination method, i.e., soldered or solderless
- Procurement cost and availability.

There are several basic types of coaxial connectors that are designed for use within a specific range of circuit frequencies and a corresponding assortment of cable styles. Some of the most commonly used coaxial connector types include the following.

A.  Type BNC.  The most popular coaxial connector for use with small-size cables are the BNC types that feature a bayonet coupling. Their electrical features include a constant 50-ohm impedance with good performance up to 8 GHz.

B.  Type TNC.  The TNC coaxial connector is physically basically the same as the BNC but with a threaded coupling. Electrically these connectors offer improved performance up to 11 GHz. They are especially useful in vibration environments where the BNC type provides unwanted radio-frequency noise.

C.  Type N.  The type N coaxial connector is popular for use with the larger cable sizes of 10 mm (0.400 inch) in diameter, although 5 to 23 mm (0.200 to 0.900 inch) sizes are available for special applications.

D.  Type UHF.  These ultra-high frequency connectors are the oldest and the least expensive of the radio-frequency coaxial connector types.

E.  Types SMA, SMB and SMC.  These subminiature coaxial cable connectors are specifically suited for use with very small flexible and semirigid cables in the size range of from 2.0 to 3.5 mm (0.80 to 0.140 inch) outer diameters. They are considered to be semi-precision high-frequency connectors that give good repeatable performance from up to 18 GHz for the SMA type, and 10 GHz for the SMB and SMC types. The type SMA has a bayonet coupling, the type SMB a push-pull coupling, and the type SMC a threaded coupling.

F.  Twinax and Triax Types.  These are high-frequency versions that are available for use in specialized applications with single-braid (twinax) and double-braid (triax) cables. They have threaded couplings and, like all of the other types, are polarized for proper positioning alignment.

Variations of several of these types of coaxial connectors are available for special applications, such as high-voltage (C, MHV, HN, etc.) and for special mountings and terminations, such as conditions under which they must be directly soldered to a printed wiring board.

## 7.5  WIRE TERMINATIONS [7]

The selection of the most appropriate wire termination method for a given application can be a very complex undertaking. However, some basic comparisons can be made.

The current trend in new product offerings shows a movement toward providing termination components and equipment to satisfy increased production requirements at a lower cost to the end product. As a consequence, electronic equipment manufacturers are gravitating toward the use of quantitatively controlled solderless devices that can be terminated individually or in a mass manner, through the use of manual and semiautomatic equipment using a minimum amount of labor. Thus, at least for electronic applications, the use of wire soldering is kept to a minimum and is avoided whenever possible. Thus, individual conductors, and those within a round multiconductor cable, are generally terminated by crimping. In most instances round-conductor flat cables are gang terminated with insulation displacement connections (IDC). Flat-conductors and fiber optic cables require the use of specialized termination devices.

### 7.5.1  Wire Crimp Terminations

The use of crimping has taken the lead in many electronic wire and cable assemblies for many electronic applications because of the ease of termination by relatively unskilled personnel. Improvements have been made in crimp technology which have the effect of removing the control (and quality) from the operator. This makes it possible to consistently make reliable terminations, regardless of the number of terminations.

Basically, the crimping method consists of tightly compressing the barrel of the terminal end of the connector contact so that it is in intimate "gas-tight" metal-to-metal contact with the stripped end of the wire within it. This can be done properly when the appropriate contact barrel size is chosen for the gauge of wire being used and when the appropriate crimping die is used.

The crimping die tooling determines the completed configuration of the termination. There are a variety of crimping tool types in use, such as those with a simple nest and indenter type die, and the more complicated four-indent die. Some tooling may work well for a given application, while for another application a different type may be superior. The major crimp configurations are shown in Figure 7.13 along with an explanation of their best application.

Many considerations affect the determination of the crimp die configuration.

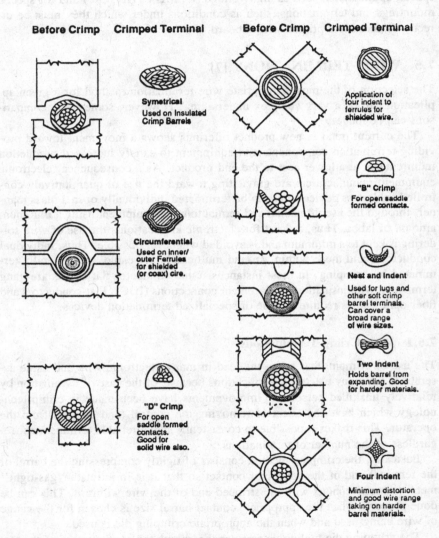

**Before Crimp     Crimped Terminal          Before Crimp     Crimped Terminal**

**Symetrical**
Used on Insulated
Crimp Barrels

**Circumferential**
Used on inner/
outer Ferrules
for shielded
(or coax) cire.

**"D" Crimp**
For open
saddle formed
contacts.
Good for
solid wire also.

Application of
four indent to
ferrules for
shielded wire.

**"B" Crimp**
For open saddle
formed contacts.

**Nest and Indent**
Used for lugs and
other soft crimp
barrel terminals.
Can cover a
broad range
of wire sizes.

**Two Indent**
Holds barrel from
expanding. Good
for harder materials.

**Four Indent**
Minimum distortion
and good wire range
taking on harder
barrel materials.

**Figure 7.13.**     Major crimp termination configurations. (*Courtesy of Burndy Corp.*)

For electronic wiring and cabling the most important are:

- The type of contact terminal to be crimped, its size, shape, material, function, and performance requirements
- The type and size range of the wires to be accommodated
- The type of tooling into which the configuration must be built
- The cost of producing the termination.

### 7.5.2 Mass Wire Insulation-Displacement Connections (IDC) [8]

The development of the U or V contact terminal opened the use of round-conductor flat cabling for a wide variety of electronic applications. The key to the cost-effective use of this termination method is the feasibility of achieving insulation piercing (displacement) as a prelude to actual wire termination in one termination step, Figure 7.14. Using ribbon cable that has been standardized to facilitate its compatibility with the IDC connector and its contacts, simultaneous multiple wire terminations can be made with a minimum of tooling and operator skill, Figure 7.11.

The U is carefully formed to produce parallel cantilever beams that allows the U to pierce the insulation easily and strip it from its locallized contact with the wire conductor. The deflected beams then force the undersize (diameter) wire into intimate (gas-tight) contact with it.

Basic IDC reliability is dependent on the high pressure that is maintained between wire and contact over extended periods of time. This ability to maintain high pressure is referred to as the spring compression reserve. The deflection

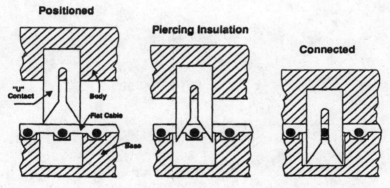

**Figure 7.14.**    Basic steps in making insulation displacement connections. (*Courtesy of 3M Electronic Products.*)

and resulting stress within the contact beams remains within the elastic and elastoplastic regions of the material, which is usually either beryllium copper or phosphor bronze. The result is that each beam continues to apply pressure at the connection throughout the life of the contact.

## References

1. "Design Guide for Electronic Wire and Cable," Belden Corporation, ECK-2-LB, 1972.
2. Arthur G. Schuh, Martin Marietta, "Fundamentals of Wiring and Cabling," *Electronic Packaging & Production*, Part 1, January 1983, pp. 242–250, Part 2, February 1983, pp. 42–28.
3. "Electronic Wire and Cable Technical Data," Brand-Rex Corporation, Technical Bulletin EC1-81, 1981.
4. "Shielded Wire and Cable," Electronic Wire and Cable Technical Data," Brand-Rex Corporation, Technical Bulletin 720, 1974.
5. Ronald A. Couch, "Miniature Coaxial Cable: State of the Art," *Electronic Packaging & Production*, February 1982.
6. "Fiber Optic Designer's Handbook," Amphenol Fiber Optic Products, F122-00188, July 1987.
7. Gerald L. Ginsberg, "Connectors: EPP Tutorial Series," *Electronic Packaging & Production*, July 1982, pp. 233–248.
8. Gerald L. Ginsberg, "IDC—a Mature Technology Continues to Grow," *Electronic Packaging & Production*, May 1983, pp. 57–60.
9. Gerald L. Ginsberg, "Connectors Link Fiber Optics to a Bright Future," *Electronic Packaging & Production*, May 1983, pp. 95–97.

# 8

# Electronic Equipment Enclosures

The packaging of the electronic equipment into an appropriate enclosure can vary from the selection of a basic off-the-shelf product to the design of a fully customized unit, Table 8.1. The high cost of designing, tooling, and fabricating these enclosures can be dramatically reduced, or even eliminated, by taking advantage of the expertise of the enclosure manufacturer. These firms offer products ranging from the outside enclosure to complete packaging systems that permit the customer to simply plug in printed board assemblies to create the basic product.

## 8.1 GENERAL ENCLOSURE SELECTION CONSIDERATIONS [1, 2]

Before any electronic equipment enclosure packaging choice can be made the following factors must be carefully weighed and assigned a relative order of importance.

### 8.1.1 Enclosure Standardization

Standards organizations throughout the world have worked to produce essentially compatible specifications. Thus, standard enclosure configurations are available that cover a variety of specifications and sizes while maintaining product uniformity, quality, and aesthetic appeal. For example, the Institute of Electrical and Electronics Engineers (IEEE 1101), the International Electrotechnical Commission (IEC 297), and the Electronics Industry Association (EIA RS-310) have standards for racks, panels, and associated equipment. However, they are

189

**TABLE 8.1.    Tradeoffs Between Custom and Off-the-Shelf Electronic Equipment Enclosures [1]**

| Design | Advantages | Disadvantages |
|---|---|---|
| Off the shelf | No tooling costs | Higher unit costs |
| | No design costs | Limited design choices |
| | Available in small quantities | Commonplace appearance |
| | Immediate delivery | |
| Custom | Unique appearance | High design costs |
| | Lower unit costs | High tooling costs |
| | Unlimited design choices | Higher minimum-order quantity required |
| | | Long initial lead time |

all generally compatible with one another and specify hardware to conform to the same basic physical outlines.

Enclosure manufacturers have benefited greatly from these and other standardization efforts and therefore they have been able to carry a large inventory of standardized enclosures and accessories. It is then the role of the packaging engineer to generate a design that incorporates these basic elements efficiently or to develop customized units that are compatible with them, unless, of course, the end product configuration is completely unique to the application.

### 8.1.2  Enclosure Cost

The cost of the electronic enclosure will, for the most part, be determined by the relative importance of the other packaging factors, but it must also be considered in its own right. The cost of a standard off-the-shelf enclosure depends primarily on the number of units purchased, the construction techniques used in its fabrication, and on its appearance. The cost of a custom-designed enclosure, on the other hand, also depends on the type of materials used, the manufacturing process, and the mechanical design. To these costs must be added the cost of the engineering effort, tooling, raw materials and fabrication labor.

### 8.1.3  Enclosure Functionality

An early packaging decision is whether or not the enclosure, or enclosure type, will be used for one or more products. If so, will it have to be produced in a variety of sizes? When an enclosure is to have multiple applications it is important to maintain as much commonality as possible in order to minimize procurement/tooling costs and to utilize common parts and components.

If the equipment is to be used with a family of products, all of the end uses must be taken into account. For example, are the end products intended to be stand alone units? If so, should they be stackable? Does the enclosure need feet or bails for tilting? Must it be rack mountable? Will it accept chassis slides? Will it accept handles? Can it be made portable? Is size important?

### 8.1.4   Electronic Equipment Environmental Stability

Protection from ambient environmental conditions, such as humidity, dust, shock and vibration, electromagnetic and radio-frequency interference (EMI/RFI), etc., are critical to sophisticated electronic systems. Obviously, enclosure packaging plays a key role in the control of these factors. The degree of protection offered in any enclosure design is often detailed in a number of national and international specifications.

Electronic enclosures can be designed to protect the equipment against precipitative dust or to be dust free. They can be drip proof or watertight.

An enclosure's shock and vibration stability, i.e. durability, is another important factor that not only determines its suitability for an application, but also for the manufacturability processes and materials to be used.

EMI/RFI considerations affect a broad spectrum of electronic applications. Consideration of this aspect of electronic packaging is necessary in order to comply with Federal Communications Commission regulations that strictly control the interference generated by electronic equipment or the equipment's susceptibility to interference generated by other products.

### 8.1.5   Electronic Equipment Enclosure Thermal Management

An ever more important consideration in equipment packaging is the enclosure's ability to handle the heat generated by the electronics within it. The degree to which this factor affects the design of the equipment is directly associated with the environmental extremes (temperature, altitude, etc.) to which the unit is exposed and its inherent ability to dissipate heat through natural conduction, convection and radiation.

Should natural thermal management means be insufficient, forced-air cooling can often be used. This then raises major air circulation concerns because of the use of fans and blowers, air intake and exhaust, and air filtration. Unfortunately, the ambient noise created by artificial cooling can be prohibitive for the end product environment.

Thermal management can also be complicated by other equipment that is near or around the unit being packaged. In these instances, assumptions sometimes have to be made concerning the effect of these other units on the actual ambient conditions. For example, if heat generating units are mounted under the unit being designed, the actual ambient temperature can be higher than that specified. Also, if another unit is mounted above the electronic product being packaged, the flow of the cooling air can be blocked or, at least reduced, to unacceptable levels.

### 8.1.6 Electronic Equipment Serviceability

Most products are intended to provide trouble free operation, but eventual failure is inevitable. Thus, when the product fails a determination has to be made as to whether or not it is to be serviceable or throw-away. When it is to be serviceable, the packaging approach must provide a sufficient level of accessibility to the assemblies to be maintained. In some applications, this need for accessibility may necessitate the use of covers, panels, captive hardware, etc., that might normally be required.

### 8.1.7 Electronic Equipment Enclosure Safety

Safety can be a key factor in the design of an electronic enclosure. This is especially true when the end product will have to meet Underwriters Laboratory (UL), International Electrotechnical Commission (IEC), or other applicable specifications. Depending on the safety features to be incorporated into the electronic equipment, special packaging provisions will have to be made for connectors, shields, markings, etc.

### 8.1.8 Enclosure Aesthetics and Quality

Electronic equipment manufacturers are becoming more concerned with the appearance and quality of the electronic packaging, especially for those units that are used for consumer applications. Is some instances, appearance and quality are closely related. For example, subtleties such as the fit of a door, the stiffness of the doors, the finish at corners and edges, and even the surface finish of the enclosure can affect a consumer's opinion of the entire product.

Poor fit or finish should be a thing of the past, as many enclosures, both off-the-shelf and custom made, are manufactured to exacting tolerances. Quality features can be built into the design ranging from stiffening ribs to multiposition latching systems. A variety of surface finishes, colors, and materials are also available.

Closely related to appearance, styling is another factor that affects the mechanical design of the equipment enclosure. This is especially true when the appearance of the equipment is intended to offer a revolutionary, rather than evolutionary, change from the older units.

## 8.2   CABINETS [2, 3]

The most common and readily available electronic equipment enclosure is the freestanding cabinet or rack. Most of these product offerings span a variety of configurations, sizes and performance characteristics. Although most cabinets accommodate internal electronic equipment units with standardized 19-inch (sometimes 24- or 30-inch) front panel widths.

### 8.2.1   Cabinet Compatibility

Many electronic equipment cabinets conform to international standards. Thus, compatibility problems can be avoided by ensuring hardware conformance to Electronics Industry Association (RS-310), International Electrotechnical Commission (IEC 297), and Deutsche Industrie Normenausschuss (DIN 41494) documents. Conformance to these standards will ensure that subracks, chassis, front panels, etc., are mounted properly and function mechanically as required.

Special attention should be paid to the basic cabinet design and its ability to be expanded into a range of cabinet sizes. The design should also remain uniform, as design conformity will dictate manufacturing cost effectiveness, and in turn, the cabinet's final price.

When a cabinet's component parts manifest proportional design consistency throughout a particular range, the cabinet with the smallest external size can provide the same standardized internal mounting dimensions as can the largest cabinet in the range. This results in an aesthetically-pleasing uniform appearance with the electronic equipment to be housed becoming more stylish in both small and large cabinets.

### 8.2.2   Type of Cabinet Construction

Electronic equipment cabinet racks are of two basic types, namely skin and modular-frame. Skin enclosures consist of side, bottom and top panels that form the overall structure. With this type of construction the internal electronic equipment units are mounted in the front of the enclosure. Thus, they are not easily adapted to changes and are not expandable for future equipment needs. However, a door can be installed at the back of the cabinet to facilitate access to the wiring and cabling between individual units and power distribution systems.

Conversely, modular-frame enclosures incorporate a frame that serves as the main structural element, Figure 8.1. Doors, panels and other accessories are then added to the modular frame to customize the enclosure.

Modular-frame enclosures offer flexibility and versatility as they can be joined together in side-by-side, one-above-the-other, front-to-back, and back-to-back

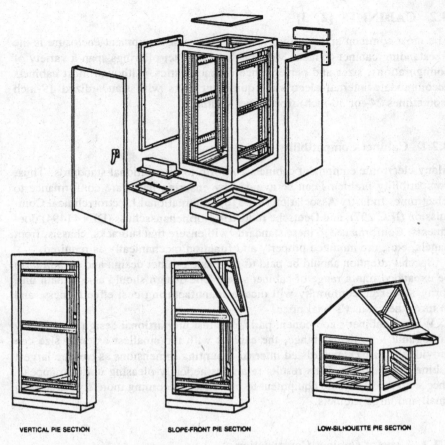

**VERTICAL PIE SECTION**            **SLOPE-FRONT PIE SECTION**            **LOW-SILHOUETTE PIE SECTION**

**FIGURE 8.1.**    Typical electronic equipment cabinet configurations. [2]

arrangements, Various shapes can also be attached to each other without the need for special adaptor panels. Thus, individually multi-enclosure configurations can be developed.

Five basic elements make up a modular frame enclosure. They are the basic frame structure, panel-mounting angles, panel-mounting support channels, cross ties, and base.

### 8.2.2.1  Basic Cabinet Frame Structure
The frame is typically made of multiformed channel that is welded together to give the frame its rigidity and durability. This type of self-supporting frame construction is one of the most economical and effective means of meeting a range of enclosure requirements.

The self-supporting construction provides both rigidity and a high capacity mounting surface while it permits the flexibility of doors and removable covers on all faces of the cabinet. The frame can be such that the internal equipment front-panel components can be recessed or mounted in swinging frames that permit access to both the front and rear of the electronic equipment mounted within the cabinet.

### 8.2.2.2  Cabinet Panel-Mounting Angles

The panel-mounting angles, typically four with vertical enclosures, are used to mount panels and electronic equipment chassis in the enclosure. The angles are recessible and removable. They are also punched for sheetmetal screws or tapped for machine screws.

The front-panel holes are usually arranged in accordance with industry standards in a repeating pattern of 0.625, 0.625, 0.500 inch spacing pattern. The mounting holes are 0.281 inches in diameter for use with clip-on nuts or are tapped with a 10-32 thread.

### 8.2.2.3  Cabinet Panel-Mounting Angle Supports

The panel-mounting angle support channels or side struts are provided to add strength to the structure. Typically made from a 16-gauge cold-rolled steel, they are vertically adjustable and removable. Their primary purpose is to support the panel-mounting angles so that mounted-equipment loads are properly distributed.

The struts can be positioned for proper load distribution, as well as for minimum interference with cabling and transverse cooling air flow. Depending on the frame height, generally from four to eight support channels are used.

### 8.2.2.4  Cabinet Cross Ties

The cross ties at the top of the frame provide strength and rigidity. They also minimize twist, sway and deflection.

### 8.2.2.5  Cabinet Mounting Base

The mounting base provides stability, and if required, convenient cabinet mobility. It can be one of many types, including pontoon, flush frame, caster frame and caster dolly.

Normally supplied with many cabinets, pontoon bases usually have a recess in the front for toe space. They are removable since they are bolted to the cabinet frame.

Flush-frame bases are structurally identical to pontoon bases, except that they fit flush with all of the sides of the frame. This is done to provide additional stability to the structure.

Caster-frame bases typically have two-inch diameter casters to give mobility to the cabinet under normal conditions. For heavy equipment loads, adverse floor surfaces, and where the center of gravity of the cabinet is high or off to one side, a caster dolly can be used that typically has four-inch casters.

Some cabinets use stabilizers. These are retractable support bars that pull out to give additional stability to the cabinet base. They are especially helpful when heavy electronic equipment chassis are extended on slides. Levelling feet also can be used to help stabilize a cabinet on an uneven floor.

### 8.2.3 Grade of Cabinet Service

Cabinet enclosures can be classified according to the grade of service, i.e., light-duty, medium-duty, and heavy-duty.

Light-duty cabinets usually house shallow, low-density (light) chassis and equipment, such as meters, indicator panels and terminals. They can be used when shock and vibration levels are minimal.

Medium-duty enclosures generally house deep, medium-density chassis and instruments. When this type of enclosure is used the shock and vibration environment should be minimal.

Heavy-duty cabinets are designed to house high-density (heavy) chassis, slide-mounted equipment, heavy power supplies, etc. They are often used when the equipment is to be exposed to severe shipping conditions and to be operated in shock and vibration environments.

### 8.2.4 Cabinet Styles

Electronic equipment cabinet enclosures are generally styled to serve various functional and operational needs. Thus, they are available as vertical racks, sloped-front cabinets, work consoles, bench-top units, open racks, and several different styles of multiple enclosures.

#### 8.2.4.1 Vertical-Rack Styling

Vertical racks are generally used when a versatile standardized enclosure is most appropriate. Typical cabinet heights provide a vertical panel (chassis) space that ranges in ten different sizes from 21 to 84 inches. Chassis/panel widths are available to accommodate 19, 24 and 30 inch units, Figure 8.2.

A modular, vertical-rack enclosure is usually chosen by:

- Selecting the required frame size
- Determining how the frame is to to enclosed, i.e., side panels, top panels, doors, etc.
- Selecting the desired panel style, i.e., plain, louvered, or grilled
- Adding accessories, i.e., shelves, drawers and casters.

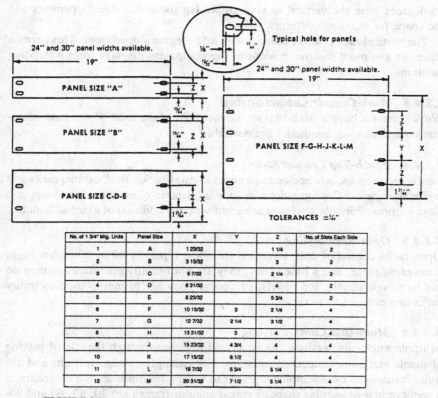

| No. of 1 3/4" Mtg. Units | Panel Size | X | Y | Z | No. of Slots Each Side |
|---|---|---|---|---|---|
| 1 | A | 1 23/32 | | 1 1/4 | 2 |
| 2 | B | 3 15/32 | | 3 | 2 |
| 3 | C | 5 7/32 | | 2 1/4 | 2 |
| 4 | D | 6 31/32 | | 4 | 2 |
| 5 | E | 8 23/32 | | 5 3/4 | 2 |
| 6 | F | 10 15/32 | 3 | 2 1/4 | 4 |
| 7 | G | 12 7/32 | 2 1/4 | 3 1/2 | 4 |
| 8 | H | 13 31/32 | 3 | 4 | 4 |
| 9 | J | 15 23/32 | 4 3/4 | 4 | 4 |
| 10 | K | 17 15/32 | 6 1/2 | 4 | 4 |
| 11 | L | 19 7/32 | 5 3/4 | 5 1/4 | 4 |
| 12 | M | 20 31/32 | 7 1/2 | 5 1/4 | 4 |

**FIGURE 8.2.**    Typical electronic equipment enclosure panel dimensions. [2]

### 8.2.4.2  Sloped-Front Cabinet Styling

Sloped-front cabinets come in three basic versions, namely:

- Sloped-front consoles
- Machine-control consoles with a work/writing shelf
- Sloped-front vertical racks with sliding shelves.

The sloped-front console is commonly used where the equipment operators are seated. In these instances the front slope is 30-degrees from the vertical to aid in the viewing of the equipment front-panels.

The machine-control console basically houses equipment that is used to automatically control the operation of manufacturing processes. For optimum viewing the front-panel slope is 30-degrees for seated operators and 45-degrees for standing operators. The lower portion of the cabinet is usually sloped inward

15-degrees from the vertical to give ample leg room for seated operators and toe space for standing operators.

The vertical rack is also available with a 15-degree sloped front. This permits an easier and more positive readout of the equipment displays from a standing position.

### 8.2.4.3  Work-Console Cabinet Styling
Work consoles have a desk-like appearance. They are used where ease of operation and pleasing aesthetics are desirable.

### 8.2.4.4  Bench-Top Cabinet Styling
Bench-top units include enclosures known as mini-racks, desk cabinet racks and instrument cases. Configurations include adjustable chassis guides, doors and sloped fronts. Portable styling is achieved with the addition of external handles.

### 8.2.4.5  Open-Rack Styling
Open racks consist of only a vertical member, typically 40 to 80 inches high, a top cross piece, and a base. Thus, they are not literally enclosures as they do not have side panels, top panels, or doors. They are primarily used as utility racks and channel relay racks.

### 8.2.4.6  Multi-Unit Cabinet Styling
Multiple enclosure sections can be joined together through the use of joining channels and interconnection hardware. For example, wedge sections and angular frames can be assembled together to form a multiple-section enclosure in a semicircular or angular shape. Typical angular frames are 30, 45, 60 and 90-degrees.

## 8.2.5  Static and Dynamic Cabinet Loading

A variety of cabinet designs are available with different mounting characteristics. Some employ high-strength extruded mounting flanges, others utilize bolt-on steel flanges.

Different manufacturers provide either static (stationary) or dynamic (movable) mounting capacity for their cabinets. The static capacity is, understandably, much higher than the dynamic capacity. In addition, if the cabinet has the potential to be moved in a loaded condition, the dynamic load capacity is of the utmost importance.

The addition of swinging frames can also adversely affect the dynamic loading characteristics. However, many cabinet manufacturers offer accessories to increase the loading capacity of the enclosure.

The design of the mounting hardware, ranging from caged nuts and bolts to screws and clip springs, affect the load characteristics of the cabinet. Thus, careful attention should be paid to the strength requirements needed to support the electronic equipment that is mounted in the cabinet.

## 8.2.6  Cabinet Accessibility

The use of doors and removable panels can facilitate having easy access to the electronic equipment housed within the cabinet. Further access can be achieved by the use of chassis slides.

### 8.2.6.1  Cabinet Doors

Doors may be installed in the front, rear, or on either side of a cabinet. They can be either flush- or surface-mounted. Lift handles, flush latches and locks are available. The doors can be installed for either left or right swing. They can also have either plain, louvered or grilled surfaces.

### 8.2.6.2  Cabinet Access Panels

Access panels and covers can be removable or hinged and fastened with screws or latches. Filler panels can be used in the front and rear of the cabinet or mounting angles in order to take up space not otherwise used by electronic equipment or doors.

Plain panels can be used to mount items such as instruments, displays, and connectors. They can also be louvered or grilled to provide for cooling air ventilation with or without the use of supplementary cooling-air filters.

### 8.2.6.3  Cabinet Chassis Slides [4]

The use of chassis slides provides a convenient drawer-type access to the electronic equipment within the cabinet for improved ease of maintenance and improved product versatility. They provide a translational motion for the load, allowing it to be locked in place in fully extended or interim locations.

Chassis on slides can also be pivoted in the extended position in order to simplify access to the top, bottom, and rear of the electronics equipment. Should the unit have to be removed from the cabinet, quick disconnect mechanisms can be used.

The wide variety of slide designs vary from simple to complex depending on the degree of sophistication required. The common slide configurations consist of two-member slides with an undertravel movement range; three-member slides with an equal or over travel range; and special slides with both front and rear travel.

The two-member slides always have an under-travel because a minimum part of the moving member has to remain within the stationary member in order to provide cantilever strength.

**TABLE 8.2.    Electronic Equipment Chassis Slide Selection Criteria [4]**

| Application | Solid bearing | Nylon roller bearing | Steel-bearing steel channel | Steel-bearing aluminum channel |
|---|---|---|---|---|
| Light load | Excellent | Excellent | Excellent | Excellent |
| Heavy load | | | Excellent | Excellent |
| Permanent | Excellent | Excellent | Excellent | Excellent |
| Quick disconnect | Excellent | Good | Excellent | Excellent |
| Vertical doors | | | Excellent | Excellent |
| Cantilevered objects | Good | | Excellent | Excellent |
| Precision mounts | | | | Good |
| Levers and links | | | Good | Excellent |
| Vibration | | Excellent | Excellent | |
| Unusual environments | | | Excellent | Excellent |
| Extremely smooth slide action | | Excellent | Good | Good |
| High temperatures | Good | | Excellent | Good |

The basic slide types, Table 8.2, are:

- Solid-bearing, metal against metal
- Plastic-roller bearing, plastic on metal
- Steel-bearing, metal and ball bearings.

A.  Solid-Bearing Slides.  Solid-bearing slides are among the lightest-load rated and least expensive types. The are usually made of fixed steel channels that slip directly over movable channels that are either zinc-plated or coated with a molybdenum disulphide finish for self-lubrication.

Depending on their size and construction, they can carry loads up to 175 pounds or more per pair with a reasonable life expectancy. Two- and three-section telescoping solid-bearing slides can, based on the length of the slide, usually extend the chassis from 10 to 30 inches.

B.  Plastic-Roller Bearing Slides.  Rugged and economical plastic-roller bearing slides are usually made with nylon rollers that can generally carry loads from 50 to 150 pounds per pair. These slides are available with many special features, such as stop-action, release arms, positive closing action to prevent drawer springback, and self-closing, if required. However, the use of nylon-roller bearing slides is limited to relatively mild temperatures. Otherwise, they have a good resistance to unusual environments and materials.

C.  Steel-Bearing Slides.  Steel-bearing slides are made with both formed steel and extruded aluminum channels. This is the most versatile type of slide since it is manufactured in many sizes and shapes. Two- and three-member telescoping steel-bearing slides give extensions in the range of from 12 to 36 inches.

The formed-steel channel types are manufactured to carry from 50 to 250 or more pounds per pair. Anodized extruded aluminum channel slides are the heaviest duty slides with a capacity of from 50 to 1,000 pounds per pair or greater when used in combinations.

**D. Special Slides.** Many diverse and unusual slides can be designed to meet specific and demanding requirements that are not met by off-the-shelf products. Thus, for example, special slides have been customized for heavy duty shock and vibration applications, and slides with very low profiles have been made that can handle upwards of 200 pounds per pair and more. Innumerable locking devices, tilting and pivoting mechanisms, finish varieties, and multistage stopping features along the length of travel have been provided.

**E. Chassis Slide Selection Criteria.** In addition to those features already enumerated, environmental considerations can play a role in the choice of a slide. For example, the solid bearing slide is unsuited for extremes of humidity or weather as its surfaces may clog or bind. On the other hand, the anodized aluminum channel slide is ideal for applications subject to extreme variations in weather and environment.

The life of the slide is related to its cycling function. It is usually unnecessary to use an expensive and durable slide when the unit it transports is rarely pulled out.

The load-carrying capacity required establishes the type of slide to be used. The length required and the amount of travel desired determines its dimensions.

The physical design of the mounting surface indicates the type or method of installation to be used. Lastly, the environment and weather conditions determine the material finishes.

### 8.2.7  Cabinet Accessories

Many accessories are available to extend the cabinet's versatility. With them, cabinets can be easily moved, associated equipment tools and cables can be stored, working/writing surfaces can be provided for convenient use, etc.

#### 8.2.7.1  Cabinet Drawers
Slide-mounted equipment drawers can be used to house electronic units, such as test equipment, that must be occasionally accessible. Enclosed storage drawers are available to contain items such as small tools, patch cords and user's/maintenance manuals.

### 8.2.7.2   Cabinet Work/Writing Shelves

Work/writing shelves can be installed at various heights within the cabinet so that they can serve as writing surfaces, work areas, or for mounting auxiliary equipment. Their use is recommended typically for loads up to 100 pounds.

Shelves can be used in light duty and limited space applications. When they are supported by the cabinet's panel-mounting angles, a shelf can slide out for use and be stored in a closed position for further use when needed.

### 8.2.7.3   Cabinet Equipment Shelves

Equipment shelves offer a way to mount heavy units such as power supplies and test equipment. Their use is ideal for equipment that is accessed periodically, but not often enough to warrant the use of slides. These shelves are designed so that they can be installed in various locations throughout the cabinet.

### 8.2.7.4   Cabinet Turrets

Special display modules in the form of turrets can be mounted in many parts of a cabinet. They are generally used to meet specific operating, space and environmental conditions. For example, turrets can offer protection from heat sources or electrical interference by placing the housed equipment away from the main enclosure. For this reason, turrets come in vertical-mount and horizontal-mount configurations.

## 8.3   SMALL EQUIPMENT ENCLOSURES [5, 6]

Choosing an enclosure for relatively small electronic units is not as formalized as is the packaging of equipment in cabinets and racks. This is mainly because plastics, in a variety of forms and compositions, are finding wide use in electronic equipment packaging. In fact, for small enclosures, Figure 8.3, plastics have some inherent advantages over the use of metals. However, a number of other factors, including the required size, quantity, finish and environment, must be considered before making a final choice.

### 8.3.1   Small Enclosure Base Material

In most electronic equipment enclosure packaging applications the metal used is aluminum. This is because aluminum provides a cost-effective compromise of properties that make it ideally suited for most housing fabrication requirements. This includes the use of sheet-metal, castings and dip brazings.

Aluminum is also used to satisfy many thermal management requirements. Although copper is a better conductor of heat, it is also heavier. Thus, the use of aluminum in heatsinks is quite common in electronic equipment.

**FIGURE 8.3.**    Typical small electronic equipment enclosures. [5]

Specifying a material for plastic enclosures, in contrast, presents a challenge. This is because there are a vast number of plastics to choose from. They vary not only in mechanical properties, but the way they are processed can significantly affect the design parameters and cost of the enclosure. The use of reinforcements, fillers and coatings also complicates the selection process.

Most plastics used in electronic equipment enclosures are thermoplastic polymers. Thus, when heated they soften enough to allow them to be forced into a mold. As they cool, the thermoplastics harden and take the shape of the mold cavity.

While many of today's electronic-grade thermoplastics are blends, it is helpful to know the basic properties of these materials. The following are some of the thermoplastic materials that are typically used in fabricating small electronic enclosures. Their selection requires an extensive trade-off analysis, as there is no single material that is best suited for most applications.

- Acrylonitrile butadiene and styrene (ABS) plastics are relatively hard and brittle. They are resistant to heat and impact, and have good low-temperatures properties and good electrical characteristics. ABS is frequently alloyed with other plastics.
- Polyamides are abrasive and impact resistant. They also offer good mechanical strength in a wide range of temperatures.
- Polycarbonates are very strong and easy to process with a predictable mold

shrinkage. They have desirable self-extinguishing properties and excellent impact strength. As with ABS, the polycarbonates are often alloyed with other plastics.

- The polyethylene materials are relatively soft, but their hardness increases with density. They are impact resistant from $-40$ to $+90°C$. and offer good resistance to most chemicals and moisture.
- The electronic-grade polyphenylene oxide (PPO) resins have good mechanical strength and toughness. They are also resistant to most liquids, steam, acids and bases.
- Polypropylenes are similar to the polyethylenes but are stronger at lower densities. A unique property of these materials is that they do not degrade with flexing. Comparatively low in cost, the polypropylenes offer good heat resistance, chemical resistance, and dimensional stability. Their mold shrinkage is both low and predictable.

### 8.3.2  Size and Quantity

Small enclosure requirements basically fall into two categories, i.e. those for prototypes and those for production. For prototypes the packaging engineer generally has no alternative but to choose an off-the-shelf product in order to avoid initial tooling costs and to speed-up the acquisition of the enclosure. Fortunately, many styles, sizes and shapes of small enclosures are readily available to satisfy the majority of prototype needs.

Naturally, if the required quantities are large and there is sufficient time, it is tempting to consider packaging the electronics in a customized enclosure. However, whether the enclosure is made of an injection-molded plastic or investment-cast metal, the typical tooling costs for a small high quality enclosure can be on the order of from $5,000 to $10,000. If die casting is used this cost can increase appreciably. Therefore, this costly initial step requires considerable confidence that the design is not likely to change significantly and that the quantity over which the tooling is to be amortized is not likely to be reduced to any great extent.

The fabrication of small enclosures from metal often requires the use of castings or the joining of piece parts by welding, riveting, or threaded fasteners. Unfortunately, the joining processes are labor intensive, and thus escalate costs. However, with careful design, metals can be used to make a less expensive enclosure than plastics for prototype and small quantity production requirements. But the intricacy of fabrication involved in working sheet metal tends to make this inadvisable for use with small electronic equipment enclosures. On the other hand, the use of sheet metal for large enclosures is often cost-effective.

The plastic injection molding and metal die-casting processes, and to a lesser extent the metal injection molding process, restricts the size of the enclosure because of the relatively limited capacities of molding/casting machines and the high design and manufacturing costs of tooling. At best, this can only be partially overcome by using structural foam materials, as the lower mold pressures involved lead to reductions in tooling costs.

The sheet-metal enclosure has no such size limitations. This makes the use of sheet-metal fabrication ideal for large enclosures, especially when the design is subject to change or modification.

As another alternative a structural foam construction can be applied to a number of plastics. This is achieved by either blowing an inert gas into the plastic or by introducing a chemical blowing agent into a plastic blend. The gas expands in the heat of processing, usually injection molding, to create a cellular structure. As a result, structural foam enclosures offer improved electrical insulation properties, light weight, and a high strength-to-weight ratio. The gas permeated materials also do not shrink when molded. This helps to avoid the appearance of sink marks on the side of the enclosure that is opposite to any ribs and bosses.

### 8.3.3  Small Enclosure Finish and Color

The finish and texture of metal enclosures are generally dependent on the process used, i.e., chemical processing, anodizing, or conventional painting. Thus, there is virtually an unlimited choice of colors and surface finishes, although paint is prone to chipping and scratching. The use of plastic-clad aluminum overcomes this drawback. The use of polyvinylchloride (PVC) or other plastic cladding generally also includes the use of textured finishes that further minimize the appearance of cosmetic defects. They also add a tough appearance to the enclosure and eliminate the need for painting or other finishes.

Plastic enclosures can have either smooth or textured finishes, or a combination of both, depending on the design of the mold. Different colors can also be selected. However, the cost can be prohibitive unless the quantity of enclosures being fabricated, approximately 1000 or more pieces, can justify the time and effort required for purging the molding machine.

The use of plastic enclosures with removable metal front and rear panels sometimes offers a good compromise. The panels can be readily punched, painted, silk-screened, etc., in a flat easy to handle manner. Molded-in bosses and alignment features, that aid in the assembly of the unit, are often provided at a minimal additional cost.

### 8.3.4  Small Enclosure Environmental Stability

A small enclosure must of necessity be required to withstand rough handling. The use of plastics, particularly the high-impact types, can often absorb an impact without permanent deformation, But, it is possible for plastics to crack. However, this can be minimized by the careful selection of the plastic (and reinforcement, if needed) and consideration of the mold design and the flow of material in the mold.

Properly gasketed metal enclosures can offer intrinsic high electrostatic discharge, electromagnetic interference, and radio-frequency interference immunity. The use of plastics in similar applications will require the costly use of an additional conductive spray coating or special polymer mix. Conductive paints can also be used, but care must be taken with surface preparation and shelf life.

It is generally accepted that it is much easier to prevent dust and moisture from getting into an enclosure, especially if the design has a tongue and groove joint between the fabricated sections of the housing. However, this is more commonly done on molded or cast enclosures, as it is expensive to provide with sheet-metal products.

## 8.4  PORTABLE CARRYING CASES [7]

While electronic equipment continues to become more complex and compact, they are also finding an increased use where portability is required. The enclosures for this type of application is generally referred to as being a carrying case, Figure 8.4.

### 8.4.1  Types of Carrying Cases

There are essentially three different types of carrying cases that can be used to satisfy various end product requirements. These essentially clamshell configurations are called transit cases, instrument cases, and combination cases.

In addition to these basic types of cases, some desktop cases may also be considered to be portable, see Section 8.3. They most often have front and rear panels, feet on their bases, and handles that facilitate their portability.

#### 8.4.1.1  Transit Cases

A transit case protects the electronic equipment while it is being transported from one place to another. The contents of the case are usually nestled in a foam cushion or suspended from some type of shock and vibration isolation device within the enclosure. This arrangement allows the imposed shock and vibration loads to be attenuated to a level that hopefully will not adversely affect the equipment.

#### 8.4.1.2  Instrument Cases

An instrument case houses a particular electronic unit while it is being operated. Thus, an instrument case is often carried within a transit case during transport when fragile equipment is being handled.

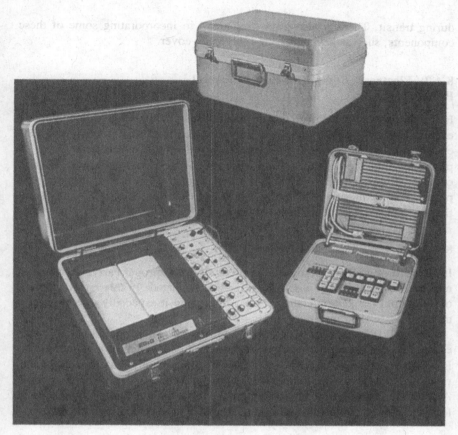

**FIGURE 8.4.**    Typical electronic equipment carrying cases. (*Courtesy of Skydyne Corp.*)

### 8.4.1.3  Combination Cases

A combination case is designed for both the transportation and operation of a piece of equipment. There are many types of combination cases, each of which is aimed at meeting the requirements imposed by a specific application. For example, MIL-T-28800 permits the use of a commercial type of combination case in moderate environments, while in another category it requires the use of a ruggedized case for more severe applications.

Some combination cases have removable covers that permit easy access to the unit during its operation. Latches are normally a part of the cover, so that there are no obstructions when the protective cover is removed. Such cases sometimes include removable covers that mount over the front and rear panels in order to protect displays and panel components that are susceptible to damage during transit. There is also a growing trend to incorporating some of these components, such as keyboards, into the panel cover.

### 8.4.2  Electronic Equipment Environmental Protection

It is important to remember that in most instances the case is operationally expendable, as its primary role is to provide protection. That is to say, its presence is not usually essential when the electronic equipment is being used. In providing this protection, a good case might necessarily become dented, crushed, or cosmetically degraded to a certain degree. In fact, if it is not, it is probably not sufficiently absorbing shock and vibration. Rather, it is transmitting this abuse to the equipment within it.

### 8.4.3  Portable Case Shell Material

In order to provide the best possible protection, the proper carrying case shell material, Table 8.3, and case design must be selected. In this regard, the use of aluminum has certain characteristics that make it especially effective when used as a case shell material.

The wall thickness for both plastic and fiberglass carrying cases must necessarily be thicker in order to equal the strength and weight of aluminum. This is because aluminum stretches and collapses more readily than the other materials, and thus it can be thinner while giving the same degree of protection. Also, in the case of deep drawn aluminum shells, the uniform wall thickness is less likely to be punctured when it buckles than is a welded section.

### 8.4.4  Portable Case Construction

A carrying case should be constructed so that shock loads placed on internal equipment can be efficiently transferred to the case. This load transfer condition is most severe on instrument and combination cases, because of the need for the control panel subassembly to be rigidly mounted to the carrying case closure frame. In these instances the shock load is transmitted from the case mounting flange through the interface joint between the case and the closure frame of the shell. Thus, the case must be able to attenuate this shock without the aid of cushions or shock mounts.

A better junction between the case and closure results in a stronger finished product. Using similar materials also helps to assure having the strongest possible case. When dissimilar metals are used it is necessary to use rivets or crimping, with some kind of sealant, to provide the joint connection. But neither riveting nor crimping is as effective as is the welding of similar metals.

**TABLE 8.3.    Portable Electronic Equipment Enclosure Material Selection Criteria [7]**

| Case material | Advantages | Disadvantages |
|---|---|---|
| Deep-drawn aluminum | Large number of sizes available from stock; high tensile strength; easy to assemble accessories by welding; wide choice of finishes available at reasonable cost; naturally corrosion resistant material; good heat conductor; provides effective EMI shielding. | It must be finished. |
| Injection molded plastic | Low cost in high volume; integral colors; wide choice of materials and colors; wide choice of styles and configurations. | High tooling costs; high setup costs; attachments must be riveted. |
| Thermoformed plastic | Lightweight; highly resistant to denting; relatively low cost tooling; wide choice of materials, including those with texture; color integral part of material. | Riveting, or other similar method, required for all attachments; poor resistance to fire. Moderate tensile strength. |
| Formed and fabricated steel | Low cost; ease of fabrication; high tensile strength; high modules of elasticity; good heat conductor; provides effective EMI shielding. | Heavy; not corrosion resistant; must be finished. |
| Fiberglass | Under light blow and normal temperature, returns to original shape after initial deformation; can be formed in one step at lower cost; extremely poor conductor; color impregnated in material. | Molds expensive, particularly match molds; considerable hand work involved in preforming and reinforcing operations; attachments must be riveted; molding process requires 2° draft in case sides; may become fragile at low temperatures, depending on resin used. |
| Wood | Low cost material; low cost tooling. | Not fire resistant, unless treated; not impervious to fungus, rodents, insects, etc., unless treated; not waterproof or water vapor resistant. |

### 8.4.5  Portable Case Closure Configuration

There are basically two carrying case closure designs. The first interlocks a male section into a female section. The male portion presses against a hollow round rubber gasket that is situated in the female cavity. The interface between the sides of the two sections transfers the shearing forces from the cover to the case when the impact is on the cover. The compression of the gasket helps to assure a watertight seal.

The second type of closure design presses the cover skin into a gasket that is retained by a z-band on the case. This gasket can either be a hollow extrusion or a molded soft elastomer. This type of closure has a basic drawback in that it does not transfer the shearing forces. Thus, external interlocking shear plates are required.

In addition, both closures are more likely to follow each other when distorted by an external force. In this way, an even amount of pressure is maintained on the gasket in order to provide a better seal.

### 8.4.6  Portable Case Heat Dissipation

It is also important to consider size, as well as strength, during the design of the carrying case. Since it is almost every electronic engineer's role to package as much electronics into the smallest enclosure, a premium is generally placed on carrying case space utilization. In addition to the packaging density challenges that this presents, there is inevitably an increased thermal management problem. Dissipating heat in a carrying case is no easy matter. Thus, satisfying these requirements may necessitate having special features built into the carrying case. It may also be necessary to alter the position of the electronics within the enclosure in order to maximum the degree of natural equipment cooling that takes place.

### 8.4.7  Weight Minimization

Weight is another important design consideration. The weight of the equipment in its carrying case will, obviously, determine whether or not it is truly portable and by whom.

In general, the weight of portable equipment should not exceed 40 pounds if one person is expected to carry it. Ideally, the electronic equipment should be much lighter. One alternative, if possible, is to divide the equipment into smaller and lighter subassemblies that can be packaged separately.

### 8.4.8  Electronic Equipment Accessibility

While packaging should always be functional, every effort should be made to ensure that the equipment is as easy as possible to operate. This not only includes considering the size and weight of the product, but also the accessibility of controls and maintainable portions of the equipment.

### 8.4.9  Shielding

As with all other enclosures, electromagnetic and radio-frequency interference requirements also apply to electronic equipment in carrying cases. Plastic cases, of course, usually will require a secondary coating with a conductive material in order to provide the necessary shielding. An alternative is to use special composite resins or fillers that have been formulated for this purpose.

Because they are electrically conductive, steel and aluminum cases provide the best inherent shielding. However, the use of metals alone will not ensure adequate interference suppression. This is because seams and other metal-to-metal interfaces, such as occur at the closure, in the carrying case, are potential shielding weaknesses. However, these weak points can be shielded with con-ductive gasketing that is an integral part of a tongue-in-groove closure interface. A lesser degree of shielding can be achieved by installing gasketing between the electronic equipment front panel and the mounting frame, because of the unshielded component-mounting openings that are usually unavoidable in the panel.

## 8.5  PRINTED BOARD ENCLOSURES [8]

Enclosures for printed wiring board assemblies in all of their various forms are the most widely used type of electronic equipment packaging enclosure. Such enclosures offer the means for tying together a unique set of printed wiring board assemblies, their backplane interconnections, and other elements of the electronic equipment.

Often referred to as card racks, card cages, card frames or subracks, they exhibit distinct differences in construction, form factor and final application.

### 8.5.1  Construction Variations

The various construction used include:

- User-assembled adjustable
- Formed and riveted or welded sheet metal
- Formed-wire frames
- Dip brazings
- Molded plastics
- Castings.

The first three construction types are often used to provide printed wiring board assemblies that are used as part of an electronic equipment system that is packaged in a cabinet rack, see Section 8.2, or similar enclosure. Thus they are generally fabricated to be compatible with the standardized chassis and panel dimensions, Figure 8.2. Conversely, the last three construction methods are generally used to make complete end product enclosures.

### 8.5.1.1  User-Assembled Printed Board Enclosures
A distinct advantage of the user-assembled printed wiring board enclosure, Figure 8.5, is that it can be supplied in a low cost kit form and assembled with threaded fasteners on an as-required basis. Because it can be readily adjusted to accommodate different printed board sizes and spacings, this type of enclosure is ideally suited for prototyping and low volume production usage.

### 8.5.1.2  Sheet-Metal Printed Board Enclosures
Sheet-metal printed wiring board enclosures are usually fabricated from aluminum or sometimes steel. The basic structure of the enclosure is usually formed by joining together the sheet-metal parts by the use of either riveting or, to a lesser extent, by welding. In general, riveting is less expensive and provides for better dimensional accuracy because its alignment is governed by precise hole placement.

Since riveted sheet-metal enclosures can be constructed faster than the threaded-fastener assembly types, their use is more cost effective for higher volume applications. Whether riveted or welded, they are also more sturdy and vibration resistant. However, once assembled, all of their dimensions are fixed and adjustments are difficult, if not impossible, to make.

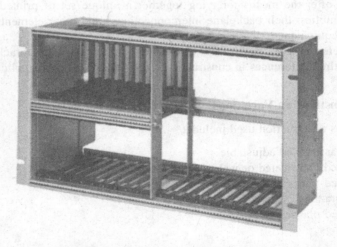

**FIGURE 8.5.**     Typical user-assembled adjustable printed wiring board assembly enclosure. [7]

### 8.5.1.3  Formed-Wire Frames

Formed-wire frames are made from heavy-gauge steel wires that are spot welded together to form a sturdy, well ventilated structure. The wires can also be formed to act as internal printed wiring board insertion and removal guides.

Wire-formed frames offer versatility of design and a relatively lighter weight as compared to the other types of construction. However, once assembled, they also are not readily modified.

### 8.5.1.4  Dip-Brazed Printed Board Enclosures

A dip-brazed printed wiring board enclosure is generally made of prefabricated aluminum parts that are permanently joined together. The dip-brazing process consists basically of first placing and interlocking the separate parts together with an aluminum filler. The assembly of the structure is then made by the melting of the filler in a vacuum furnace (fluxless vacuum dip brazing) or by immersion in a molten brazing flux. The filler metal flows to create brazed joints between interfacing parts in a manner similar to the soldering of butt or lapp joints.

Dip brazing allows for the joining of both thick- and thin-walled parts without distortion because, unlike welding, the heating is uniform throughout the entire structure. In addition, the filler metal also can reach and flow into generally inacessible areas, thereby sealing all interfaces for maximum shielding and water tightness.

### 8.5.1.5  Molded Plastic Enclosures

Injection molding can also be used to make printed wiring board enclosures for special applications. When light weight and rigidity are required, this type of enclosure can be made with a foamed plastic. However, relatively thick walls are required in order to provide the structure with adequate strength. Special provisions have to be made if shielding is required. This type of construction is cost-effective only for high volume applications. This is because of the relatively high investment required in the specialized tooling needed for each different end-product configuration. However, the price per piece is usually low if the tooling and setup costs are amortized over a sufficiently large number of enclosures.

### 8.5.1.6  Printed Board Casting Enclosures

Castings can be used when a rugged one-piece printed wiring board assembly enclosure is needed. The selection of the appropriate casting method is a trade-off between piece price and tooling/setup amortization. Thus, the use of investment castings has the highest unit price and the lowest tooling cost; die casting has the lowest unit price and the highest tooling cost; sand casting are somewhere between these extremes. Sand castings also have a surface roughness that requires the machining of critical dimensions after casting.

## References

1. Norbert Laengrich, Racal-Dana Instruments Inc., "Designing the Ideal Product Enclosure," *Electronic Products*, June 30, 1981, pp. 47–49.
2. Frederick J. Duquette, Schroff Inc., "Considerations for Selecting Electronic Cabinets," *Electronic Manufacturing*, Part 1, September 1987, pp. 45–47; Part 2, October 1987, pp. 30–31.
3. Paul J. Dvorak, Staff Editor, "Modular Enclosures: Beyond the Basic Box," *Machine Design*, November 6, 1986, pp. 101–106.
4. Joseph Del Vecchio, Grant Hardware Co., "Chassis Slides: A Component for Today," *Electronic Packaging & Production*, February 1976, pp. E19–E22.
5. Miles Leadbetter, Vero Electronics Inc., "Selection Guide—Part One: Small Enclosures," *Electronic Products*, August 1980, pp. 51–53.
6. Susan Crum, Assistant Managing Editor, "Options in Enclosure Materials Abound," *Electronic Packaging & Production*, December 1985, pp. 58–59.
7. Paul Johnson, Zero Corporation, "Sizing Cases for Portable Equipment," *Electronic Products*, August 18, 1982, pp. 77–80.
8. Howard Markstein, Western Editor, "Card Cages Fundamental in PC Board Packaging," *Electronic Packaging & Production*, November 1985, pp. 98–101.

# 9

## Thermal Management

The thermal management of electronic equipment encompasses all of the natural and artificial processes and technologies that can be used to remove and transport heat from individual components in a controlled manner to the ultimate system heat sink, Figure 9.1. In general, it is not usually too difficult to package the equipment in such a manner that it can operate properly at a nominal end-use thermal enviroment. The challenge arises in attempting to provide the electronics with stability in an environment with variations in storage and operating temperature, not to mention the effects on the environment of other heat-generating (or airflow blocking) units in or around the electronics being packaged [1, 2].

A primary objective of thermal management is to ensure that all circuit components, especially the integrated circuits, are maintained within both their functional and maximum allowable limits. The functional temperature limits provide the ambient or component package (case) temperature range within which the electronic circuits can be expected to meet their specified performance. However, it is the temperature of the heat-generating source within the component package, e.g., the integrated-circuit junction, that must ultimately be maintained with its allowable temperature range. Thus, the component manufacturer must define a parameter (theta) that relates the rise in junction temperature with respect to the component case or ambient temperature per unit amount of power dissipation. Operation outside this temperature range may be expected to result in excessive equipment performance degradation. Thus, if overall system reliability objectives are to be met, the thermal management must ensure equipment operating stability within acceptable limits when the assembly is exposed to the specified (or actual) enviromental variations. Other secondary objectives include ensuring that:

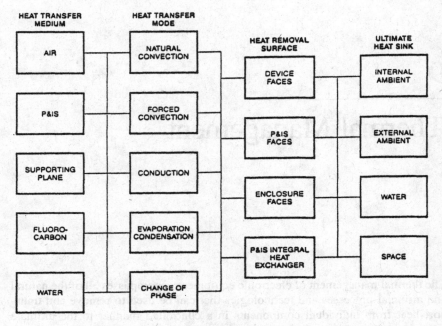

**FIGURE 9.1.**    Thermal management cooling options (IPC-PD-335).

- Exposed surfaces are kept within safe temperature limits
- Overall subsystem cooling meets availability and reliability objectives
- Cooling system design is consistent with the ambient heat dissipation capability
- Cooling system cost is appropriate for the end product cost.

All of these objectives must be kept in mind throughout the electronic equipment packaging process. Thus, as with many design processes, the optimum thermal management approach will be the result of a series of technical choices and tradeoffs.

## 9.1  BASIC THERMAL MANAGEMENT CONSIDERATIONS [3]

The dissipation of the heat generated within electronic equipment results from the interaction of the three basic modes of heat transfer, i.e., conduction, convection, and radiation. These heat transfer modes can, and often do, act simultaneously. Thus, any thermal management approach should attempt to maximize their natural interaction. (Where very high dissipation levels are encountered, special heat transfer modes are sometimes employed that are based on the use of either liquid cooling or vaporization (ebullition) cooling.)

### 9.1.1 Thermal Conduction

The first mode of heat transfer to be encountered is that of conduction which takes place to varying degrees through all materials. The conduction of heat through a material is directly proportional to the thermal conductivity of the material ($K$), Table 9.1, the cross-sectional area of the path flow, and the temperature differential across the material; it is inversely proportional to the length of the path, i.e., the thickness of the material. (The variation of the $K$ value with temperature can be neglected within the limits of the enviromental conditions that are relevant to electronic equipment packaging.)

The following rules apply in order to optimize the transfer of heat by conductivity:

- Use materials that have the highest thermal conductivity that is consistent with structure, fabrication, price and availability considerations.
- Utilize an optimum cross-sectional area.
- Maintain the temperature differential at as low a value as possible.
- Keep the thermal path as short as possible.

TABLE 9.1.  Heat Sink Material Thermal Conductivities [9]

| Material | K(BTU/hr/ft$^2$/°F/ft) | K(Watts/Inch-°C) |
|---|---|---|
| Still air | 0.016 | 0.0007 |
| Alumina (99.5%) | 16 | 0.70 |
| Beryllia (99.5%) | 114 | 5.00 |
| Silver | 242 | 10.6 |
| Diamond | 364 | 16.0 |
| Gold | 172 | 7.57 |
| Epoxy | 0.114 | 0.005 |
| "Thermally conductive" epoxy | 0.45 | 0.02 |
| Aluminum alloy 1100 | 128 | 5.63 |
| Aluminum alloy 3003 | 111 | 4.88 |
| Aluminum alloy 5052 | 80 | 3.52 |
| Aluminum alloy 6061 | 99 | 4.36 |
| Aluminum alloy 6063 | 111 | 4.88 |
| Copper alloy 110 | 226 | 9.94 |
| Beryllium copper 172 | 62–75 | 2.7–3.3 |
| Brass alloy 360 | 67 | 2.95 |
| Stainless steel 321 | 9.3 | 0.41 |
| Stainless steel 430 | 15.1 | 0.66 |
| Steel, low carbon C1040 | 27 | 1.19 |
| Titanium | 4–11.5 | 0.2–0.5 |

Copper has the best practical thermal conductivity of the available heat transfer materials. However, because of copper's weight, cost, and ease of fabrication, aluminum is the thermal management material most frequently used in electronic equipment.

In the case of aluminum, care should be taken in the selection of an appropriate alloy, since conductivity is affected directly by the alloying element. However, other factors, such as formability, strength, environmental resistance and cost must also be considered.

Epoxy, even thermally-conductive epoxy, has relatively poor conduction characteristics. Thus, except for special applications, its use should be avoided when more efficient means of heat transfer can be used.

## 9.1.2  Thermal Radiation

Thermal radiation is the transfer of heat by electromagnetic radiation, primarily in the infrared (IR) wavelengths. It is the only means of heat transfer between bodies that are separated by a complete vacuum as would happen in space environments.

Heat transfer by radiation is more difficult to analyze than heat transfer by conduction. The rate of heat flow by radiation is a function of the surface of the hot (heat-dissipating) body with respect to, among other things, its emissivity, Table 9.2, its effective surface area, and the differential of the fourth power of their absolute temperatures.

The emissivity is a derating factor for surfaces that are not black-bodies. It is defined as the ratio of emissive power of a given body to that of a black body

TABLE 9.2.    Normal Emissivity of Various
Surfaces [9]

| Material and finish | Emissivity |
|---|---|
| Aluminum sheet—polished | 0.040 |
| Aluminum sheet—rough | 0.055 |
| Anodized aluminum—any color | 0.80 |
| Brass—commercial | 0.040 |
| Copper—commercial | 0.030 |
| Copper—machined | 0.072 |
| Steel—rolled sheet | 0.55 |
| Steel—oxidized | 0.657 |
| Nickel plate—dull finish | 0.11 |
| Silver | 0.022 |
| Tin | 0.043 |
| Oil paints—any color | 0.92-.96 |
| Lacquer—any color | 0.80-.95 |

for which emissivity is unity. It should be noted that the term "emissivity" has little to do with color in the optical sense; bodies of any optical color can have high emissivities and be referred to as being normal black-bodies. For example, the emissivity of anodized aluminum is the same if it is black, red, or blue. However, its surface condition is significant; a matte or dull surface finish will be more radiation efficient than will be a bright, glossy surface.

The actual surface area of the hot body must usually be derated by a shielding factor in order to yield an effective surface area. In many designs, extrusions in particular, the radiant energy emitted by one heatsink fin is often reabsorbed by an adjacent fin because radiant energy flows on a path that is perpendicular to the surface. This energy flow will continue until it is absorbed by air molecules or by the surface of another (cold) body. Thus, closely spaced fins have little effective radiation area because the radiant energy merely bounces back and forth from fin to fin. In these situations "less is more," that is to say, having fewer fins (spaced further apart) is more efficient than having more fins. This should be also be kept in mind when components are placed within the equipment or on printed wiring board assemblies. In this situation, when a high-temperature component is located near a heat dissipator, the heat dissipator will absorb some energy from the other component, causing it to be hotter than it normally would be. With these considerations in mind, the thermal management approach can generally take maximum advantage of radiation by:

- Using heat sinks that afford maximum surface area
- Keeping surface finishes as highly emissive as possible
- Using heatsink materials that will conduct the thermal energy to the dissipating surfaces at as fast a rate as possible in order to keep the temperature differential between the dissipating surface and air or the absorbing surface as high as possible.

### 9.1.3  Thermal Convection

The convection heat transfer mode is the most complex. It involves the transfer of heat by the mixing of fluids, usually air.

The rate of heat flow by convection from a body to a fluid is a function of the surface area of the body, the temperature differential, the velocity of the fluid and certain properties of fluid.

The contact of any fluid with a hotter surface reduces the density of the fluid, and thus causes it to rise. The circulation resulting from this phenomenon is known as free or natural convection. The air flow can be induced in this manner or by some external artificial forcing device, such as a fan or a blower.

Heat transfer by forced convection can be as much as ten times more efficient than natural convection, Figure 9.2. However, its use should be minimized

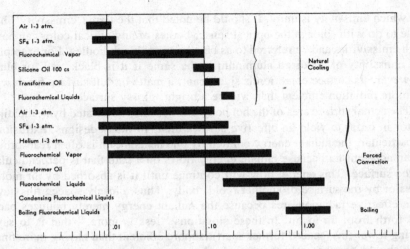

**FIGURE 9.2.**    Heat transfer coefficients for various coolants and modes of convection. [1]

because the use of such heat transfer devices can be costly, power consuming, space consuming, noisy, and most importantly, can sometimes affect equipment reliability.

Other application factors that should be considered in order to maximize the advantages of convective cooling include:

- Position a heat-emitting surface so that its larger dimension is in the vertical plane.
- When applicable, mount heat-generating components as low as possible on its heat sink.
- Provide proper enclosure ventilation so as not to impede the natural convection flow of cooling air.
- Position components within the equipment such that the more heat sensitive components are closer in the cooling airflow intake than are the less sensitive components.
- In forced convection keep the lower power components upstream and the higher power components downstream.
- Provide proper air flow ducts (channels) within the equipment so that optimum pressure heads can be maintained, thereby insuring higher air flow velocities, and thus more efficient heat transfer.

### 9.1.4   Altitude and Temperature Effects

Convection and radiation are the principle means by which heat is transferred to the ambient air. At sea level approximately 70% of the heat dissipated from electronic equipment might be through convection and only 30% by radiation. However, as altitude increases, the convection factor becomes less effective as the air becomes less dense, so that at 70,000 feet the heat dissipated by radiation might increase to as much as 70% or more. Obviously, this can be a critical factor when packaging equipment for aircraft installations.

Since air density is directly proportional to air temperature, convective heat transfer efficiency is reduced at higher ambient temperatures by a thinning out of the air molecules. As in the case of altitude, adjustment to anticipated convective cooling should be takn into account if the electronic equipment is expected to operate over an extreme range of ambient temperatures.

At ambient temperatures up to approximately 50°C. there is usually no need to derate the convective cooling values derived at 25°C. but the effect of the increased ambient temperature must still be directly applied when determining semiconductor case or junction temperatures.

## 9.2  GENERAL THERMAL ANALYSIS [4, 5]

The thermal analysis should be performed as early as possible in the overall equipment packaging process. Usually, the first and best opportunity is during the conceptual phase, when it is possible to estimate component operating temperatures. This is also the best time to try to uncover severe problems, if any, with the thermal aspects of the thermal management approach being taken. A failure to do so can only result in later increases in implementation costs and schedule delays.

As a prerequisite to any thermal analysis, it is important that the design team establish overall power dissipation limits at this time for the equipment and all of its critical elements. This will serve to place constraints on the electrical designers and get them to consider thermal management in an effective manner when they develop the circuitry and partition it into subassemblies. It is also important that all affected design disciplines take into account that the design should not be overly sensitive to small increases in power dissipation or ambient temperature. In addition, it is also necessary to define the maximum steady-state conditions (temperature) at which the equipment is expected to operate reliably. It is also necessary to know the maximum allowable junction temperatures and heat dissipating parameters of the critical circuit components.

The overall packaging concept should be evaluated before any detailed calculations are made. This entails examining:

- The type of cooling to be used at various locations within and around the equipment, including internal and external cooling paths
- The overall unit size, including external surface size, type and orientation
- Subassembly type, size, quantity and orientation.

With this information and the rough power-dissipation estimates, it will be possible to initially select the degree to which natural and artificial cooling methods, Figure 9.3, will be incorporated in the thermal management approach. If the packaging approach seems feasible, a more detailed worst-case thermal analysis is needed to ascertain whether or not individual maximum component junction temperatures are exceeded.

### 9.2.1  Component-Level Thermal Analysis

Heat generated within a semiconductor component is basically transported by thermal conduction to the external surfaces of the package where it is then transferred to an external heat sink by conduction or to cooling fluids, usually air, by a convection process. Although complex analytical and numerical models are often required to simulate these processes, a convenient way to characterize the results is the terms of internal and external thermal resistances, Figure 9.4. The internal thermal resistances represent the temperature rise per unit of power dissipation that will occur within the component package; the external resistances represent the temperature rise per unit of power dissipation that will

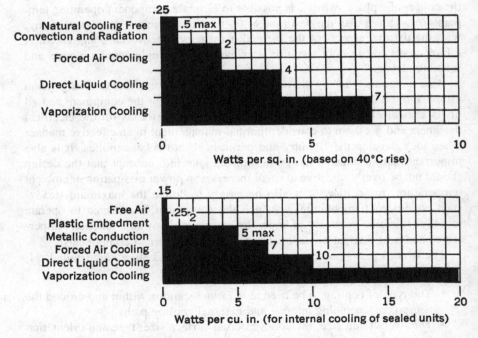

FIGURE 9.3.    Cooling method comparison. [2]

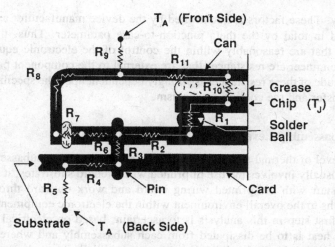

**FIGURE 9.4.**    Cross-section of a typical air-cooled electronic component. See Figure 9.5. [1]

occur between the external surfaces of the package and the next level of cooling. For an air-cooled component, the combination of these parameters can be represented by a parallel resistive circuit, Figure 9.5. (If a heat sink is used, the external resistances can be modified accordingly.) This circuit can be used to determine acceptable values of internal and external resistance for given values of power dissipation and junction temperature, or vice versa. As is readily seen, the primary parameters that determine junction temperature are coolant or heat-sink temperature, internal thermal resistance, external thermal resistance, and the amount of power being dissipated. The amount of power will, of course, be determined by the circuit technology and equipment architecture which, as such, represents a parameter that the thermal management approach can do little to modify or control. The magnitude of the internal resistances, on the other hand, are determined largely by the package geometry, materials, and method

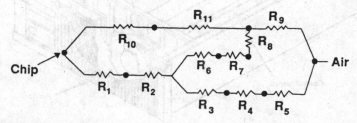

**FIGURE 9.5.**    Thermal analog of a typical air-cooled electronic component. See Figure 9.4. [1]

of assembly. These factors are controlled by the device manufacturer and usually defined in total by the theta junction-to-case parameter. Thus, the only parameters that are reasonably within the control of the electronic equipment packaging engineer are resistances that are external to the component package. The magnitude of these resistances is strongly dependent upon the specific mode of heat transfer and the cooling mechanism.

### 9.2.2  Subassembly-Level Thermal Analysis [6]

The next level of thermal analysis applies to circuit component subassemblies. Since this usually involves the use of printed wiring board substrates, it is often easiest to start with the printed wiring board and work outward through its cooling paths to the overall environment within the electronic equipment enclosure. The first step in this analysis is to ascertain, based on electrical inputs, how much heat is to be dissipated from each subassembly and where on the board the larger heat-dissipating components are initially located. Depending on the results of the final analysis, it may be necessary to modify the location of the components board and the subassemblies within the enclosure in order to optimize the heat dissipation. How the heat is dissipated on the various printed wiring board assemblies and within the enclosure will depend on the mode of thermal management being employed. In most applications direct convection cooling is used that is either free (natural) or forced (artificial), Figure 9.6. In

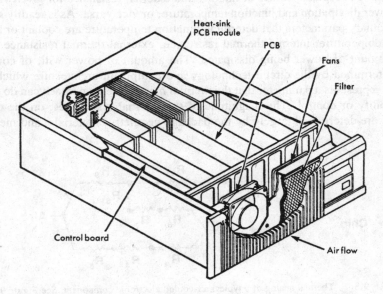

**FIGURE 9.6.**  Typical forced-air cooled electronic equipment. [6]

some instances, the efficiency of these approaches is enhanced by the use of heat sinks on the larger heat dissipating components in order to minimize hot spot problems.

### 9.2.2.1  Subassembly Convective Heat Transfer

The dependence of convective heat transfer on the local air velocity is well known and is consistent with the layman's everyday experience. Many formulas—formulas that have been tested for their performance in electronic equipment cooling applications—exist for the computation of convective heat transfer parameters. All of these formulas show a dependence on both air velocity and component geometry.

A thermal analysis should consider several elements that are relevant to a printed wiring board assembly in an enclosure. The first requirement is to determine both the total volumetric airflow and the airflow distribution. In this regard, it is important to keep in mind that filters, grills, card guides, and other mechanical packaging hardware will restrict the airflow. Even if the airflow is found to be adequate, the desired printed wiring board assembly thermal analysis will require boundary conditions that consist of the air temperature at the board air inlet and the airflow in the associated assembly channel. Fortunately, there are reasonably adequate airflow and air temperature prediction techniques available to the electronic equipment packaging engineer. Thus, calculators and personal computers can be used to determine these values.

The air velocity distribution over the assembly affects two issues:

- The local air temperature in the vicinity of any given component
- Convective heat transfer from component and board surfaces to local air.

Having an unrestricted inlet and exit of the flowing air greatly simplifies the air temperature computational problem. The situation is much more complex to analyze when consideration is given to the effects of the actual three-dimensional geometry of the complete printed wiring board assembly. In these instances the airflow and air temperature computational problem is rather difficult, if not impractical, to solve due to both the cost and time considerations.

The analysis of a component mounted in a printed wiring board is far more complex than is the analysis for a single part by itself. This is because many different component sizes and shapes are usually found in a single assembly. Also, intra-component voids are common and irregular. These complex geometries tend to void any standard textbook convective heat transfer correlations for very simple shapes. Thus, they are tyically valid for only one surface, as opposed to the many surfaces of a printed wiring board assembly.

### 9.2.2.2  Subassembly Conductive Heat Transfer

Contributions by conductive heat transfer from components to the printed wiring board and within the printed wiring board itself depend on component mounting style and the amount of board metallization. For example, components mounted in sockets conduct the least amount of heat to the board, soldered through-hole mounted components conduct a little more heat, and surface-mounted components with many terminals transfer the greatest percentage of their heat to the board. Naturally, special heat coupling techniques between the component and the board would be an even more efficient conductive heat-transfer mechanism. However, nothing is ever free, because printed wiring board conduction that cools one component may heat up one or more of its neighboring components. Therefore, the actual interaction of conductive heating effects on all of the board-mounted components must be a part of the thermal analysis.

Fortunately, of all of the physical mechanisms involved in heat transfer, conduction is the best understood. The partial differential equations are straightforward, the computational algorithms for their solution have been explored extensively, and there are many excellent programs available to aid in the analysis. The difficulty in applying them to printed wiring board assembly applications is that the geometry of the metallization patterns for the many layers and holes is very complex. This geometric complexity is not outside the capability of computer-aided computational methods, but just the size of a typical problem can be overwhelming.

### 9.2.2.3  Subassembly Radiative Heat Transfer

Radiative heat transfer at the printed wiring board assembly level is probably a negligible consideration in forced-air cooling applications. However, sealed or vented enclosures have such small air movement that in these situations, radiative heat transfer should be considered, particularly between component surfaces and the opposing printed wiring board or structural surfaces.

The major complexity in radiation computations at this level is the fact that shielding effects by neighboring components must be considered. In addition, there is significant "bouncing around" or radiation reflection that takes place that must be evaluated.

### 9.2.2.4  Subassembly Thermal Evaluation

A common printed wiring board assembly thermal analysis evaluation procedure consists of testing an assembly using either thermocouples attached to selected components or scanning with an infrared imaging system. In either case, the discovery of excessive temperatures often results in a redesign of at least a portion of the assembly. It also requires the fabrication of the assembly prior to the analysis.

The ideal situation would be to have an analysis procedure that accurately predicts board and component temperatures prior to fabrication. This need is recognized by computer-aided design (CAD) vendors who offer thermal analysis programs to implement the basic printed wiring board assembly design package.

The basic physics of the conduction and radiation are well understood. The conduction element is, however, a challenging computational issue. The convection is also difficult to compute accurately. Thus, when selecting a printed wiring board assembly thermal analysis program the following should be considered:

- How detailed are the conduction calculations? What board geometry and layout detail is considered?
- Does the program provide a method to estimate the volumetric airflow entering the board area (channel) or must it be guessed?
- Does the program consider airflow that enters the assembly inlet area in a nonuniform manner?
- Do the convection calculations consider the interaction of components and missing components? Is a flat board (without components) asumed?
- Are radiative effects from components considered? How are shielding effects evaluated? Are radiation reflections from neighboring components and adjacent assemblies considered?
- What computational method is the program based on, i.e., finite difference or finite element?
- Can the claims of program accuracy be accurately backed up?
- Was the program developed especially for electronic thermal analysis applications?
- Can the program be tailored for specific applications, such as cold-plate attachment, conduction cooling to side rails, etc.?

It is important to keep in mind that the thermal analysis of a printed wiring board assembly in an enclosure with other heat-generating assemblies is very complicated and will necessitate the making of many assumptions. Thus, in the long run, this aspect of thermal management will not be finally resolved until an actual piece of end product equipment is analyzed.

### 9.2.3 Enclosure-Level Thermal Analysis

For thermal management purposes an electronic equipment housing can be considered to be either of the enclosed types, in which the inside of the unit is completely separated from the outside ambient environment, or the ventilated type, in which the inside air is supplied from the outside enviroment and subsequently returned to it in a heated condition.

Heat from within an enclosed housing is dissipated indirectly. That is to say, the internal heat must first be transferred to the external cooling surfaces of the equipment (or to a cold plate) through the use of internal conduction, natural convection and radiation, and subsequent conduction through the walls of the enclosure. In some applications, such as military airborne equipment, this form of thermal management is mandated in order to exclude outside (abrasive/polluted) air from entering the enclosure.

In the case of the more common ventilated electronic equipment packaging enclosure configuratin the ambient cooling air can be introduced into the enclo-

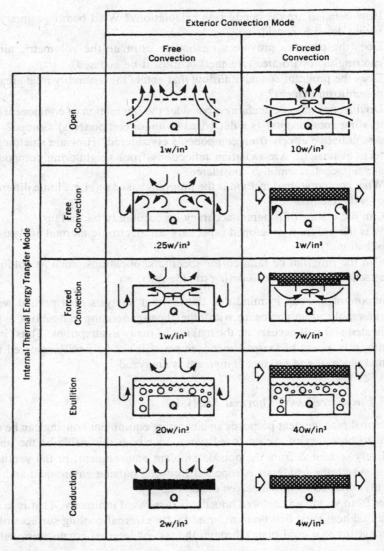

**FIGURE 9.7.**    Various heat transfer capabilities (40°C. temperature differential. [3]

sure by means of natural (free) ventilation or forced convection, Figure 9.6. In this manner, the heat generated within the enclosure is dissipated to the ultimate ambient heat sink by the heating of the cooling air as it passes through the unit.

Figure 9.7 illustrates some typical thermal performance levels that can be obtained for different enclosure configurations. As is easily seen, proper enclosure ventilation can greatly enhance dissipation capabilities in both natural and forced convection modes.

### 9.2.4  System-Level Thermal Analysis [7]

The thermal analysis of an entire system can be quite extensive and complicated. The project paths in Figure 9.8 show the wide-ranging inputs that might be required at to implement the packaging of a high-performance electronic system in a cabinet enclosure. The success of such an undertaking can be optimized by taking a few basic precautions, such as:

- Using conservative, but reasonable, assumptions throughout the electronic equipment packaging activity
- Early modeling of the design
- Having continual cooperation and communication among the personnel responsible for logic/circuit design, power-system design, hardware packaging and thermal management.

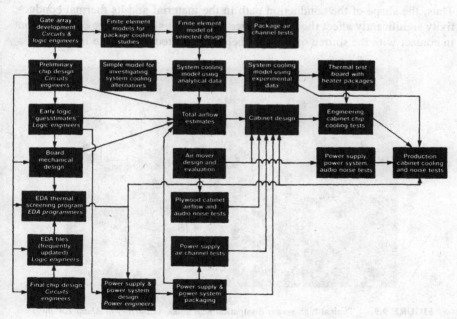

**FIGURE 9.8.**    Typical thermal management design project. [7]

To some extent these activities are mutually dependent. For example, conservative assumptions are usually made because the input information is always qualified as to what is unknown and what the probable bounds of error are. Also, early modeling can be possible when effective cooperation is provided. Thus, it is important that the personnel not directly responsible for packaging and thermal management realize the importance of these activities so that they continually supply information that might affect the thermal performance of the system. Another benefit of this approach is that logic/circuit design and debugging can be performed on a stable equipment design without interruptions for fixes or delays caused by errors that might be attributable to cooling problems.

## 9.3  HEAT SINKS [8, 9]

A good heat sink tends to minimize temperature gradients and maximize surface area per unit volume consistent with the cooling technique employed. Thus, heat generated from a component at one point on the heat sink is readily conducted throughout the heat sink to its exposed convection-cooled surfaces, Figure 9.9.

### 9.3.1  Heat Sink Shapes

The smaller the temperature drop across the heat sink, the more efficient it is. Thus, the shape of the conduction path in the material and its thermal conductivity significantly affect the performance of the heat sink. A large cross-section in contact with the source of the heat being dissipated aids in the heat flow.

**FIGURE 9.9.**    Typical high-power dissipation heat sinks. (*Courtesy of Aham Tor Inc.*)

The heat sink's shape exposes the maximum surface area that is consistent with the mode of cooling employed. Natural convection cooling imposes limitations on heat sink fin spacing. This is because when the fins are closer together than approximately 10mm (0.4 inch) the natural convection mode of heat transfer becomes impaired. Conversely, the use of forced convection allows the fins to be spaced closer together and even have them serrated in order to increase surface area.

Some heat sinks employ fins at angles to each other to minimize heat reflection between fins, while other geometries use parallel fins in order to maintain a rectangular profile. Still other configurations promote cooling fluid turbulence and special arrangements permit air flow from any direction. Other considerations in the shape of heat sinks are component attachment methods and other packaging requirements.

### 9.3.2  Heat Sink Materials

Heat sink materials are selected for their ability to perform as extended surfaces. Of course, the materials must also be good thermal conductors Table 9.1.

Silver is the best metal conductor, but it is much too expensive for most applications. Copper is an excellent thermal conductor that finds many applications where high conductivity is required. Steel is used in high-pressure applications and other areas where high strength is important.

Aluminum is the most widely used heat sink material for electronic packaging applications. This is because its ease of fabrication and relatively light weight (less than one-third the specific gravity of copper) make it an excellent material for forming long individual fins. It is also approximately one-half the cost of copper in a raw material size that conducts the same amount of heat. Unfortunately, aluminum is difficult to join by soldering or brazing, but it is very machinable.

### 9.3.3  Fabricated Heat Sink Types

The fabrication technique dictates the shape of the heat sink possibly more than do any thermal considerations. Each manufacturing method and material has certain capabilities and limitations that determine the final shape.

#### 9.3.3.1  Extruded Heat Sinks

An aluminum extrusion with a finned cross-section, constitutes a convenient raw material from which to fabricate heat sinks. The fins on such a heat sink must necessarily run parallel to the length of the original stock.

There are limitations to the configurations of the extruded fins such that the height of adjacent fins cannot, as a rule, be longer that four times the width of the gap at the narrowest point between the fins. Exceeding this ratio increases the risk of tool breakage. However, this ratio does not apply to adjacent non-parallel fins.

Many ingenious shapes have been extruded. Thus, an extruded heat sink is usually selected from a manufacturer's stock of sizes, configurations, and shapes.

Extruded heat sinks may be quite large, so that they can also be used as structural members of the electronic equipment. In fact, the trend is to design enclosures with the heat sink as a main supporting element. Often the heat sink comprises most, if not all, of the entire rear panel of an enclosure. Heat sinks can also be relatively small, so that they can be individually pressed onto semiconductor devices in small transistor-outline "TO-" metal cans or with molded plastic shapes.

Extrusions are usually preferred for mounting several devices on one heat sink. However, they are limited in that air can flow only along the length of the extrusion. Some packaging arrangements take advantage of this by using the heat sink to form an air channel through the equipment. Fans are then mounted at one end of the channel to force the cooling air through the passage.

Extruded shapes may be formed with hollow passages for liquid coolant flow. Such liquid-cooled heat sinks are often used when thermal densities exceed ten watts per cubic inch.

### 9.3.3.2  Stamped Heat Sinks

Fabricating heat sinks from metal is a simple and economical way to manufacture low- and medium-power heat dissipators. Press-on heat sinks for the popular TO-5, TO-18 and TO-92 case styles come in a variety of stamped configurations.

Many stamped heat sinks employ vertical fingers formed at an angle to the component mounting base. These finger allow air to flow freely through the heat sink from various directions. Some variations lance out splines from an other wise flat sheet in order to form protuberances to promote turbulence in moderate airflows, they also enhance heat transfer by radiation. Still other arrangements form fingers in a staggered array. All of these configurations are used to maximize the space available for heat transfer.

### 9.3.3.3  Die Cast Heat Sinks

Heat sinks can be die cast when the quantity to be made is sufficient to offset the relatively high tooling costs. These are several advantages to using die-cast heat sinks:

- They may be used in the same application as extruded heat sinks without being limited by fin orientation
- Boxes, cavities, holes and other features may be formed in the casing
- Smooth contours, textured surfaces and rounded edges can be formed.

A drawback, in addition to cost, is the limited flexibility of die casting tooling. For example, provisions must be made for mold ejector pads on the part, and it is expensive to change the design during a production run. However, within limits, it is possible to machine the die cast (or extruded) shape to provide some secondary features or to make small changes to the heat sink.

A variation of die casting is impact extrusion whereby the part is formed in a mold in a plastic state under extreme pressure. Here again the initial investment is high, but the piece price becomes quite reasonable in large quantities.

### 9.3.3.4  Pin-Fin Heat Sinks

An array of cylindrical or conical splines bonded or otherwise attached to a base plate is probably the most efficient metal-to-air heat transfer arrangement. This geometry accepts airflow in the plane of the array or perpendicular to it. Heat sinks incorporating such an arrangement are particularly adapted to situations in which a fan blows directly down onto the fins. The air then spreads around the pins while it passes them. Unfortunately such heat sinks are not easy to fabricate, although some die cast versions have been made.

### 9.3.3.5  Machined Heat Sinks

Quite a variety of custom heat sink configurations have been machined from solid plate or bar stock. Some have one or more radial fins that form an extended surface around a cavity within which the component is retained. Other screw-machine products incorporate threaded fasteners that improve heat transfer between the component and the heat sink or between the heat sink and the next assembly, such as a chassis or printed wiring board.

As previously mentioned, it is not uncommon to machine basic heat sink extrusions or die castings to provide additional features or features that cannot be formed by the original fabrication process. In fact, performing secondary machining operations is a general practice within the industry. So is the providing of anodizing, chemical surface finishes, etc.

## 9.3.4  Heat Sink Design

The first and most important step in heat sink design is to obtain a realistic estimate of the heat load. At times, this appears to be the most elusive parameter to define. Once this is accomplished, careful consideration must be given to the heat sink operating mode, Figure 9.10.

### 9.3.4.1  Natural Convection Without Radiation Cooling

Mode 1 involves using the heat sink with only natural convection without taking advantage of radiation cooling effects. A typical Mode 1 application occurs when the heat sink is located in an enclosure, without any drafts, and the finned

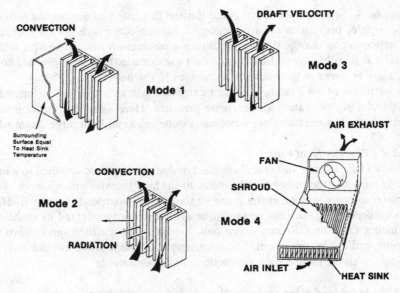

**FIGURE 9.10.**    Heat sink operating modes. [9]

surfaces are facing other parts of the electronic equipment whose operating temperature is comparable to that of the heat sink.

In such cases it is imperative that the packaging engineer fully understand the implications of what heat transfer mechanisms are, and are not, taking place. Of particular importance is the need to significantly derate the heat dissipation capabilities of the heat sink, since they were probably determined using analytical techniques that assumed that the heat sink was able to interact with its environment without any impediments.

### 9.3.4.2  Natural Convection With Radiation Cooling
Mode 2 involves using the heat sink with both natural convection and radiation cooling effects. This is the preferred natural heat transfer arrangement. Within reason, this mode of operation should not require a derating of the originally calculated performance capabilities of the heat sink. Such calculations are generally based on the following assumptions:

- Fin height to fin spacing ratio of from 1.0 to 4.0
- Maximum heat sink length of approximately 0.15 meters (6 inches)
- Sea level atmospheric pressure
- Ambient temperature of 21°C.

Such assumptions are necessary in order to assure that there is no pinching of the air boundary layer between adjacent fins. Otherwise a natural convection heat transfer coefficient correction factor must be used to represent that actual end-product environment in which the heat sink operates.

### 9.3.4.3  Unshrouded Draft Cooling

Mode 3 involves using the heat sink primarily in an unshrouded configuration with draft convection cooling, with or without radiation effects. This mode of operation occurs when an unshrouded heat sink is installed in a system that provides moving air, of a known and consistent velocity, over its finned surfaces. With this mode of heat transfer it is conservative to assume that only natural convective cooling takes place. Therefore it is more cost effective to perform calculations based on the available draft velocity adjacent to the fins. This thermal management approach is based on integrating the heat sink into a much larger system that provides the moving air naturally by having proper ventilation, artifically using air movers (fans or blowers), or a combination of both. If the free ventilation draft air flow is sufficient it is sometimes possible for the equipment to operate naturally or during an air-mover failure with some degradation of heat transfer in a soft failure mode.

This cooling approach is extremely valid for unmanned equipment that must continue to operate for long periods of time between maintenance actions when the heat sink performance is based on relatively low draft velocities. Obviously, when the heat sink is designed for high draft velocities this failure mode does not apply.

### 9.3.4.4  Shrouded Forced-Convection (Cold Plate) Cooling

The design of a forced-convection shrouded heat sink (cold plate), Mode 4, involves having the cooling air either pushed or pulled through the airflow area. Whether the air stream is pushed or pulled, the finned surfaces offer resistance. Therefore, the total cooling package performance must take into account the pressure drop caused by this resistance in order to minimize air-mover size, input power, and cost.

Optimum heat sink performance is not directly related to whether the air is pushed or pulled through the finned passages. The criteria should be based on packaging, maximizing airflow distribution, air mover life, transition ducting simplicity, and cosmetic appearance.

Some forced-convection heat sink design procedures assume that there is a nonconducting shroud configuration, i.e., the shroud's primary function is to enclose the fins and form an airflow passage duct. Generally, it is cost effective not to require thermal continuity between the shroud and the fin edges. However, in marginal designs, this continuity provides an additional design safety factor. Also, when high-power components are mounted on a shroud, it is mandatory that the shroud be conductively and thermally coupled to the fin edges by brazing or by the use of thermal conducting compounds or mounting pads.

### 9.3.4.5  General Heat Sink Design Considerations

Most extruded aluminum heat sinks have relatively thick fins and webs that tend to thermally equalize (isothermalize) its mounting surfaces. Thus, when the heat is uniformly distributed the thermal gradient across the heat sink is usually much smaller than the inlet-to-outlet air temperature rise. Since the inlet air is cooler than the heat sink discharge air, it might be expected that the heat sink surface temperature will track the air temperature profile. However, in most cases it does not. The heat sink isothermalization occurs because of axial heat conduction through the fins and web cross-section in a heat flow direction that is opposite to the air flow.

Uniformly loaded extruded aluminum heat sinks exhibit almost uniform temperature profiles, i.e., the inlet temperature droop is minimum. Thus, the thinner the heat sink cross-section, the larger the thermal droop.

There are no simple rules of thumb for estimating heat sink thermal droop. If such an estimate is required, a computer-aided thermal analysis should be performed, or the tentative thermal management arrangement should be tested.

The majority of heat sink hot spots occur not by accident, but by design oversight. Thus component placement should be based on trying to achieve heat (power) dissipation uniformity, as well as ease of packaging and assembly.

## 9.4  AIR MOVERS (FANS AND BLOWERS) [10, 11]

If it has been determined that forced-air cooling is appropriate for the electronic equipment being packaged care must be taken in selecting the appropriate air mover (fan or blower) for the application. Thus, the packaging engineer must be concerned with heat transfer efficiency, the compactness of the cooling paths, and the noise generated by the air-moving device. At the subassembly level this means taking care not only of the printed wiring board design but also of the method of attaching the subassembly to the enclosure, as these factors contribute significantly to heat transfer efficiency. At the system level this means, among other things, positioning parts and subassemblies in such a way as to optimize air flow, i.e., minimize the number of turns and obstructions in the airflow path. For example, the spacing between subassemblies should be sufficient to minimize resistance to the airflow, and thus allow the airflow rate to be maintained. However, having too large a spacing allows the airflow across the components to drop, thereby jeopardizing heat-transfer efficiency.

Effective conductive heat transfer on subassemblies can also help to minimize the demands on the forced-air cooling arrangement so as to help minimize the size of the required fan or blower. Placing the major heat-dissipating elements of the system so that they are directly accessible to the cooling air flow will also help in this regard.

Once the equipment is configured to move air effectively, the next objective is to select an air mover that will run efficiently at the system's operating point.

### 9.4.1  Air Mover Operating Point

The motion of air through electronic equipment can only be accomplished by the creation of a pressure drop across the unit in the same manner that current can be caused to flow through a resistor by the application of a voltage across it. This resistance to flow is sometimes called the system impedance.

Due to the complex nature of the flow path typicall found in electronic equipment, pressure drop ($\Delta P$) calculations do not easily yield to the customary fluid flow equations. In these instances, most packaging engineers find it best to resort to testing the equipment in order to determine its airflow impedance characteristics.

When a fan is applied to force air through electronic equipment, the flow through the system is determined by the intersection of the performance curve and the impedance curve, Figure 9.11. At this point of operation, the pressure available from the fan to force air through the system is equal to the pressure required by the system for that amount of flow. If a qreater flow is required, a fan of higher pressure characteristics will be required in the same manner that a power source of higher voltage is required to increase the current through a resistor.

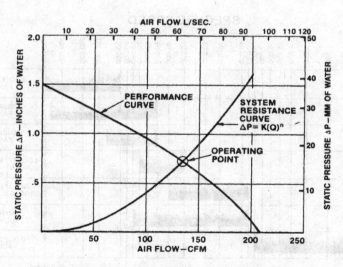

**FIGURE 9.11.**    Air mover operating point determination. [11]

### 9.4.2  Fan Laws

When a fan delivers air into a fixed system, the effects of changes in air density and speed can be determined by the methods of dimensional analysis. Such an anlysis can be instructive in showing the enormous power penalty to be paid by increasing air flow by increasing the speed of the fan. For instance, to effect a 20% increase in air flow, the speed of the fan would similarly have to increase by 20%, the static pressure would rise by 44% and the horsepower would rise by 73%.

### 9.4.3  Fan Specific Speed

The basic application factors that must be known before attempting to select the best type of air-moving device for a specific application are air flow rate, static pressure, and the speed (or speeds) at which the fan may operate. By the methods of dimensional analysis, these factors may be used to yield a figure of merit called specific speed.

There is a range of specific speed for which each generic type of air-moving device is best suited by means of efficiency. The air-moving devices that yield the highest efficiency in delivering high rates of air flow against low static pressures will have high specific speeds. Conversely, those air-moving devices best suited for low flow rates at high pressures will have low specific speeds. Figure 9.12 shows the range of specific speeds for the types of air-moving devices commonly used in electronic equipment packaging applications.

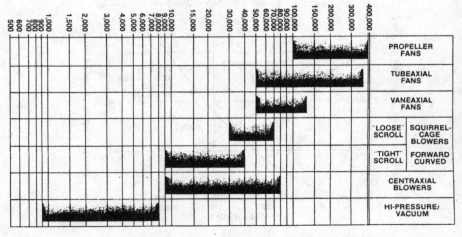

**FIGURE 9.12.**    Air mover specific speed. [11]

### 9.4.4  Moving Air Acoustical Noise

All air-moving devices produce acoustical noise. However, though careful aerodynamic and motor design, this acoustical noise can be held to a minimum consistent with the airflow and pressure capabilities of the air mover. For the lowest acoustical noise in a given application the following steps should be taken:

- Component placement and airflow paths should be carefully analyzed in order to assure that the least amount of air is required to achieve proper heat transfer.
- Unnecessary restrictions to air flow that increase the impedance of the system should be eliminated.
- The correct air-moving device should be selected on the basis of specific speed.

To further decrease the acoustical noise output it is sometimes advisable to use sound-absorbing materials, such as sheets of plastic foam, to cover the air-moving components. The sheets can also be applied to the interior surfaces of enclosures and ducts. However, care should be taken in the use of these materials as they are often heat insulators. As such, they will cause the ambient temperature to rise within the equipment. Moreover, the use of foam material can increase the airflow resistance of the system. Unfortunately, these effects may require the use of a larger (noisier) air mover in order to maintain the appropriate levels of heat transfer within the enclosure.

If interior space is available, an expansion chamber can be used as a muffler to reduce transmitted noise. Another alternative is to isolate equipment air entry and air exit points from the system operator because these locations are often the sources of the unacceptable noise. In addition, louvers and grills should be selected for low pressure drops in order to allow for a smooth airflow pattern throughout the system. Louvers over front-panel air inlets can also serve as sound barriers.

### 9.4.5  Air-Mover Selection

Propeller and tubeaxial fans are the most widely used for electronic equipment cooling because of the low-pressure conditions found in most applications. While vaneaxial fans have higher peak static efficiencies, they are more expensive and tend to be noisier because of their stationary flow-straightening vanes. A propeller fan is less expensive than a tubeaxial device, but it is also less efficient.

In the high-pressure cooling systems on minicomputers and microcomputers the tightly packaged components produce higher heat loads and higher systems resistances than in most other electronic units. Thus, the lower specific speeds of centrifugal blowers make them more efficient than fans in these applications. Moreover, backward-curved wheel units have higher peak static efficiency than forward-curved blowers. Thus, while backward-curved types find their way into critical low-noise applications, forward-curved units sometimes produce a quieter system. This is because the higher-pressure capability of the latter allows for lower operating speed, and thus lower noise.

## 9.5  LIQUID COOLING [12, 13]

Higher circuit speeds and high-density packaging significantly increase the heat density in electronic equipment. Needless to say, this complicates the use of conventional thermal management techniques. Thus, the use of forced-air cooling is increasing, and in some instances is giving way to liquid cooling. The applications for liquid cooling include large-scale computer systems, military radar, telecommunications equipment, medical imaging and power supplies.

Liquid cooling for most electronic equipment packaging applications can be divided into two basic technologies:

- Direct immersion of the electronics in a cooling liquid, and
- Indirect cooling through the use of a liquid-filled cold plate, heatsink blanket, or heat pipe.

### 9.5.1  Cooling Liquids

The coolant for these applications must have special physical, chemical, and electrical characteristics in order to serve this function. Thus, the typical types of fluids used to cool electronic equipment are fluorocarbon liquids. In addition to the properties listed in Table 9.3, these inert dielectric coolants have other desirable properties such as:

- A high dielectric constant (approximately 1.9 at 1 kHz)
- A very-low surface tension of 15 dynes per sq. cm. (versus 72 for water)
- Nonflammability
- High thermal stability
- Low chemical reactivity.

Perfluorinated liquids contain no undesirable hydrogen or chlorine. Their fluorine molecule component gives them thermal and chemical stability with low solvent power. They are also compatible with metals, plastics, and elastomers, and are resistant to the effects of corona discharge in both liquid and vapor form.

A free-convection heat-transfer coefficient for a typical electronic-grade cooling liquid might be in the range of 250 W/m/K (about 25 times greater than that for air). Heat transfer rates for a perfluorinated coolant under forced-

TABLE 9.3.    Heat Transfer Properties of a Typical Electronic Cooling Liquid [13]

| | Liquid room temperature (25°C) | Liquid at boiling point (97°C) | Vapor at boiling point (97°C 1ATM) |
|---|---|---|---|
| Density | | | |
| kg/m$^3$ | 1778 | 1602 | 13.6 |
| lb/ft$^3$ | 111 | 100 | 0.85 |
| Thermal conductivity | | | |
| W/m/K | 0.064 | 0.057 | 0.014 |
| BTU/(h)(°F/ft) | 0.037 | 0.033 | 0.008 |
| Specific heat | | | |
| J/kg/K | 1047 | 1172 | 963 |
| BTU/(lb)(°F) | 0.25 | 0.28 | 0.23 |
| Dynamic viscosity | | | |
| kg/m/s | 0.00142 | 0.00046 | 0.00002 |
| centipoise | 1.42 | 0.46 | 0.02 |
| Coefficient of thermal expansion | | | |
| °C$^{-1}$ | 0.0014 | 0.00046 | 0.0027 |
| °F$^{-1}$ | 0.0008 | 0.0009 | 0.0015 |

convection conditions might be well over 1,000 W/m/K (about 50 times greater than that for air).

The heat-transfer characteristics of perfluorinated liquids do not approach those of water. However, their inertness and dielectric properties allow their use in direct contact with electronic components and interconnection wiring with absolutely no effect on circuit operation.

The fluid for liquid-cooled cold plates can be either a refrigerant (as part of a refrigeration system), chilled water, or tap water. Ester silicate and ethylene glycol/water mixtures are also used with cold plates because they exhibit low viscosity, good thermal characteristics, compatibility with aluminum and copper, and offer performance durability.

## 9.5.2    Direct Liquid Cooling

Direct immersion cooling can be considered to be the ultimate in liquid cooling. This is because it takes advantage of the superior heat-transfer characteristics of liquids by bringing them into direct contact with heat-generating components in the electronic equipment.

In general, direct liquid cooling offers a more efficient means of thermal management than does the use of forced air. It also offers local packaging efficiency with regard to the size of the electronic assembly being cooled because fans, blowers, ducts, and air filters are not required as an integral part of the equipment enclosure. However, liquid immersion cooling can increase the size of the overall system because of the need for external pumps, heat exchangers, and holding tanks. A fully- or partially-sealed enclosure is also required. Another consideration is the cost of the fluid. For example, fluorocarbon liquids can cost as much as $1,000 per gallon.

There are three basic operational modes for dielectric liquid cooling of electronic equipment. The fluid can be either static for heat transfer by means of the natural convection or nucleate boiling, or it can be pumped in order to provide a fluid flow for forced-convection cooling, Figure 9.13.

Just as with natural air cooling, free-convection direct liquid cooling is based on the reduction of fluid density that occurs when the fluid is warmed as it comes into contact with the hotter heat-dissipating components of the equipment. However, unlike natural air cooling, a fixed volume of liquid coolant, equal to the unused spaced within the sealed enclosure, is used. The fluid rises as it is warmed so that it displaces the cooler (heavier) fluid above it. The warmed circulating liquid will also come into contact with the walls of the electronic enclosure. Thus, as long as the walls of the enclosure are maintained at a temperature that is less than that of the cooling fluid, the fluid will be lowered in temperature so that it can return to be heated again, and thus naturally continue the circulation process.

Forced convection liquid cooling increases heat transfer efficiency over that of natural immersion liquid cooling because it is able to use a larger volume of coolant. The efficiency can also be increased by moving the fluid at a greater speed across the surfaces of the heat-dissipating components. However, it does require the use of an external circulating pump. The external heat sink may be ambient air, a thermal transfer heat exchanger/cooler, or refrigeration equipment.

Under some circumstances the most efficient immersion cooling method uses nucleate boiling. This heat-transfer offers the lowest thermal-resistance characteristics, and thus the highest potential for component cooling.

The efficiency of nucleate-boiling heat transfer is contingent on the formation of bubbles on the heat-dissipating surfaces. The bubbles are formed as the surface temperature reaches a certain level (boiling incipience) so that agitation

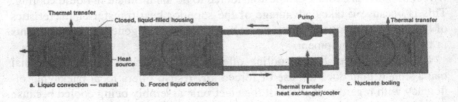

**FIGURE 9.13.**    Modes of liquid immersion cooling. [13]

efficiently transfers the heat from the component to the coolant. In some applications, the hot bubbles are then cooled when they come into contact with cooler heatsink fins that extend into the space just above the surface of the liquid source pool in the enclosure. The condensate drips back into the pool to continue the cooling cycle.

Nucleate boiling has an advantage in that the local heat transfer that takes place can be up to ten times greater than that of natural-convection cooling alone. However, a disadvantage is the need for thermal overshoot, whereby an initially higher temperature is required to create the first bubbles.

### 9.5.3  Liquid-Cooled Cold Plates

A common method of indirect electronic equipment liquid cooling occurs with the use of cold plates. Such devices operate by the principle of conduction-coupled cooling. In this mode of heat transfer the heat dissipated by circuit components is conducted to a metal plate that is cooled by a flowing liquid. If the heat transfer is accomplished by other means, such as liquid-to-liquid or liquid-to-air, the system is then referred to as being a heat exchanger. Heat exchangers can also be considered as a liquid-cooling technique. However, they are more often used as an adjunct to the primary cooling system, Figure 9.13b.

A liquid- or air-cooled cold plate is often configured as a flat metal plate over serpentine tubing for fluid flow. It can also be in the form of a machined or cast metal block that has an internal cavity or channels, plus inlet and outlet ports. Internal fins can also be included for more effective heat transfer. The parts of the cold plate are usually dip or vacuum brazed together. For maximum cooling efficiency it is recommended that circuitry be mounted on both sides of the cold plate. Ideally, the heat load should be apportioned equally to both sides of the assembly. If need be, the surfaces of the cold plate can be insulated from the circuitry. However, doing so will degrade the heat-transfer performance of the assembly. Therefore, the space between the circuitry and the plate should be kept as small as possible, and the material used in the space between them should be as thermally conductive as possible.

The cooling capacity of cold plates depends upon the surface area of the cold plate, the temperature of the coolant, coolant flow rate, and the materials and fluids used. For example, a basic cold plate design with copper tubing sandwiched between two 97 × 155 mm (3.8 × 6.1 inch) interlocking aluminum plates can dissipate 2 kW with an inlet water temperature of 40°C.

Surfaces as large as 150 × 585 mm (6 × 24 inch) can offer a thermal resistance of 0.01°C./W from plate surface to coolant at a flow rate of 1 gpm. For maximum performance all-copper cold plates having convoluted fins within the water passage can also achieve a thermal resistance of 0.01°C./W for a 150 × 150 mm (6 × 6 inch) cold plate at a flow rate of 1.5 gpm.

### 9.5.4  Liquid-Filled Heat Sink Blankets [14]

A newly emerging method of indirect electronic equipment liquid cooling is the use of liquid-filled heat sink (LHS) blankets. Such devices consist of a fluorocarbon fluid in a self-contained, easy-to-install, multilayer film package. The thin and flexible blanket must conform to the contours of the components on a printed wiring assembly. The heat is then conducted from the surface of the components to an external surface of the blanket. This, in turn, heats the liquid within the blanket. Natural convection then takes place within the blanket in order to transfer the heat to the enclosure chassis, cold plate, heat spreader, or heat exchanger, Figure 9.14.

The natural convection of the liquid results in an averaging of temperatures between devices and provides a more uniform temperature gradient across the assembly. This isothermal effect permits a given assembly to operate at a higher clock speed. The advantage of this approach is that the LHS blanket is completely passive and requires no external accessories. It has no moving parts, is rugged and puncture resistant. Problems of dust contamination, fan noise, air/fluid flow restriction, and fan failure can be eliminated with this approach to thermal management. Assembles and components cooled in this manner also benefit from mechanical cushioning. This protective effect is particularly significant when it is used with electromechanical components or other heat-dissipating components that are physically larger than integrated circuits.

Printed wiring board assemblies with up to 2.5 W/cm$^2$ (16 W/in.$^2$) are potential candidates for this mode of cooling. Higher power densities may be more effectively handled by means of direct immersion cooling.

### 9.5.5  Heat Pipes [15]

Other new methods are being used to dissipate the heat generated in electronic equipment. One such approach uses the heat pipe, a device that can move thermal energy from one location to another. It can accept heat at high thermal densities and reject it at low thermal densities, or vice versa, with a very low temperature drop. This can be done more efficiently than a solid-metal conductor and without any moving parts. Specifically, heat pipes have been shown to have three significant features:

- Heat-pipe heat exchangers offer significant improvement in thermal performance when compared to the use of solid-aluminum heat exchangers.
- Performance improvements can be obtained with a simple and maintainable thermal interface between module and chassis sidewalls.
- The reliability of circuit components increases significantly when a heat-pipe module is used as compared to using an aluminum plate, especially at elevated coolant temperatures.

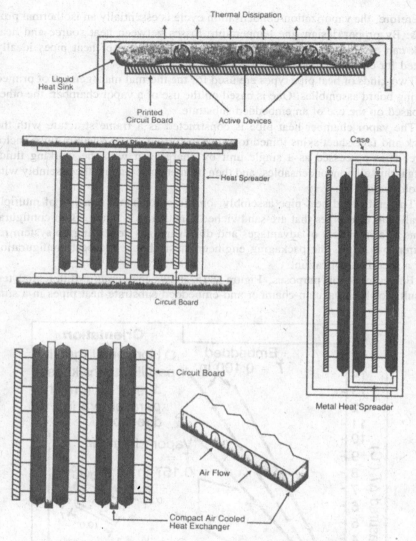

**FIGURE 9.14.**   Liquid heat sink blanket design options. [14]

Essentially a heat pipe is a closed container with a capillary wick structure and a small amount of vaporizable fluid. It employs a vaporization/condensation cycle whereby the capillary wick pumps the condensate to the evaporator (heat source) at which point the working fluid evaporates and removes heat. The vapor pressure drop between the evaporator and the condenser is very small.

Therefore, the vaporization/condensation cycle is essentially an isothermal process. By proper design, the temperature losses between heat source and heat sink can also be made very small. This makes the use of heat pipes ideally suited for electronic component cooling applications.

Two kinds of heat pipe types are used for the thermal management of printed wiring board assemblies. One is based on the use of a vapor chamber, the other is based on the use of an embedded substrate.

The vapor chamber heat pipe is constructed as a frame structure with the wick and face sheet skins joined together to form a single vacuum-tight enclosure. It is processed as a single unit by charging it with the working fluid, outgassing all noncondensables, and then hermetically sealing the assembly with a cold weld.

The embedded heat-pipe assembly, on the other hand, consists of multiple small, flat heat pipes that are sandwiched together in a frame. Both configurations offer a variety of advantages and disadvantages depending on system requirements. Thus the packaging engineer must choose the best configuration for a specific application.

For comparison purposes, Figure 9.15 shows temperature differential test results for both vacuum-chamber and embedded substrate heat pipes in a sim-

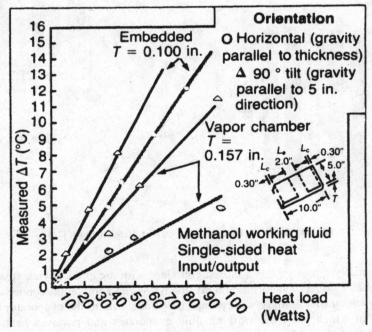

**FIGURE 9.15.**    Printed wiring board heat pipe performance comparison. [15]

ulated avionics chassis application. When tested for maximum flux limitation, the individual embedded 1.5 mm (0.060 inch) thick heat pipes have been shown to transport 35 watts in a horizontal position at a heat flux loading of 3.6 W/cm before reaching the beginning of dryout. Vapor-chamber heat pipes of a similar design have reached a heat flux of 5 W/sq. cm.

When tested for thermal transport, a 2.4 mm (0.094 inch) thick embedded printed wiring board assembly heat pipe substrate transferred 200 watts. A similar 4.0 mm (1.57 inch) thick vapor-chamber heat pipe transported 500 watts.

## 9.6 THERMOELECTRIC HEAT PUMP MODULES [16]

Some electronic equipment packaging engineers are turning to the use of solid-state thermoelectric heat pump modules for unique thermal hot spot management applications. Typical modules ranging in size from 2.5 to 50 mm (0.1 to 2.0 inches) square and from 2.5 to 6.3 mm (0.1 to 0.25 inches) thick can chill a heat dissipating component down to, and substantially below, ambient temperatures.

A basic thermoelectric principle is that when a closed circuit is made from two dissimilar metals, an electrical current (DC) flows when the junctions are maintained at different temperatures. Reversing this process (the Peltier effect), a thermoelectric heat pump can be created by inputting a current so that a temperature differential is maintained between two solid-state device junctions.

Thermoelectric modules for electronic applications are made using bismuthtelluride doped to create the semiconductor (N- and P-type) elements. The couples, connected in series electrically and in parallel thermally, are integrated into modular devices. The modules are packaged between ceramic plates with a high mechanical strength in compression for optimum electrical insulation and thermal conduction.

In a typical electronic component cooling application, the hot side of the thermoelectric heat pump module is attached to a heat sink or other heat transfer element. The cold side is mated to the component to be cooled. When a direct current is applied, electrons pass from a low energy level in the P-type material to a higher energy level in the N-type material. This causes heat to be removed from the cold-side ceramic plate.

Single thermoelectric modules can achieve temperature differentials of up to 65°C. and can pump tens of watts of heat. Greater differentials can be achieved by stacking one module on top of another, a process known as staggering or cascading. Two modules stacked together can achieve temperature differentials of 85°C. or more; three- and four-stage devices can attain up to 105°C. and 125°C. respectively. With these capabilities, the use of thermoelectric heat pump modules is attractive in electronic packaging applications where heat must be dissipated in the range of from a few milliwatts to 100 or more watts. Typical applications include military and aerospace assemblies, fiber-optic and other communications systems, medical laboratory equipment, and commercial, industrial and consumer products.

# References

1. R. C. Chu, R. E. Simmons, IBM Corporation, "Thermal Management of Large Scale Digital Computers," *International Journal for Hybrid Microelectronics*, Volume 7, Number 3, September 1984, pp. 35–43.
2. Michael W. Gust, The Mitre Corporation, "Thermal Considerations in System Design," Proceedings, *International Electronic Packaging Society Conference*, 1987, pp. 10–25.
3. International Electronic Research Corporation, "Heat Sink/Dissipator Products and Thermal Management Guide."
4. Kinishi Itoh, "Heat Management Faces Demand of High Thermal Density," *Electronic Packaging & Production*, January 1987, pp. 136–139.
5. S. Witzman, K. Graham, Bell-Northern Research and Northern Telecom Electronics Ltd., "Design Considerations for Cooling in Telecommunications Equipment," *Proceedings International Electronic Packaging Society Conference*, 1987, pp. 1036–1047.
6. Gordon N. Ellison, Tektronix Inc., "The PCB Thermal Analysis Problem," *Printed Circuit Design*, October 1987, pp. 27–30.
7. E. A. Wilson, Honeywell Bull Inc., "Thermal Management Design for a Large Mainframe," *Electronic Packaging & Production*, February 1989, pp. 142–144.
8. Jack Spoor, Aham Tor Inc., "Heat Sink Application Handbook," 1974, pp. 1–35.
9. Joel Newberger, Aham Tor Inc., "Thermal System Approach to Heat Sink Selection," 1979.
10. Carlos C. Chardon, Torin Corp., "Design Equipment to Run Silent, Run Cool," Reprinted with permission from *Electronic Design*, (vol. 30, No. 6), June 21, 1980, pp. 119–125.
11. Rotron Inc., "Selecting/Specifying Fans and Blowers," ENG 010A, March 1978.
12. Howard W. Markstein, "Liquid Cooling Optimizes Heat Transfer," *Electronic Packaging & Production*, April 1988, pp. 46–49.
13. Tom Dewey, "Developments in Liquid Immersion Cooling," *Electronic Packaging & Production*, April 1988, pp. 59–61.
14. 3M Industrial Chemical Products Division, "Fluorinert™ Liquid Heat Sink Technical Description and Application Data," 98-0211-6032-4, February 1989.
15. A. Basiulius, C. P. Minning, Hughes Aircraft Co., "Improving Circuit Reliability with Heat Pipes," *Electronic Packaging & Production*, September 1986, pp. 104–105.
16. Milton Levine, Melcor/Materials Electronic Products Inc., "Solid State Cooling with Thermoelectrics," *Electronic Packaging & Production*, November 1989, pp. 74–77.

# 10

# End Product Applications

The end-product applications for electronic equipment packaging span a wide range of products that vary from high-volume/low-cost consumer electronics to low-volume/high-cost space-age technology. Therefore, as can be expected, each end product configuration and packaging implementation is unique to the application. However, although the results vary considerably, the underlying considerations to be taken into account generally fall into several basic categories. For example, Table 10.1 shows the basic worst-case environments associated with nine electronic equipment categories. Thus, the following information is intended to provide examples of some of the general concerns that are associated with the packaging of several different categories of equipment. However, it is important to keep in mind that considerations discussed for one category of electronic product can often be appropriate for one or more of the others.

## 10.1 CONSUMER ELECTRONICS [1]

A typical example of the sophisticated products being made available in consumer electronics is the handheld calculator, Figure 10.1. Using the productivity advantages provided by the use of computer-aided design and manufacturing (CAD/CAM) tools, the packaging of this type of electronic equipment is initially based on design-for-manufacturability considerations. As will be shown, this attention to manufacturability in the initial phases of the development of electronic equipment packaging designs was well worth the effort.

TABLE 10.1. Realistic Worst-Case Electronic Equipment Use Environments (*Courtesy of the Institute for Interconnecting and Packaging Electronic Circuits.*)

| Use category | Environment | | | | | |
|---|---|---|---|---|---|---|
| | $T_{min}$ °C | $T_{max}$ °C | $\Delta T^*$ °C | $t_D$ hrs | Cycles/ year | Years of service |
| Consumer | 0 | +60 | 35 | 12 | 365 | 1-3 |
| Computers | +15 | +60 | 20 | 2 | 1460 | ~5 |
| Telecomm | -40 | +85 | 35 | 12 | 365 | 7-20 |
| Commercial aircraft | -55 | +95 | 20 | 2 | 3000 | ~10 |
| Industrial and automotive passenger compartment | -55 | +65 | 20 &40 &60 &80 | 12 | 185 100 60 20 | ~10 |
| Military ground and ship | -55 | +95 | 40 &60 | 12 | 100 265 | ~5 |
| Space | -40 | +85 | 35 | 1 12 | 8760 365 | 5-20 |
| Military avionics | -55 | +95 | 40 to 80 &20 | 2 | 500 | ~5 |
| Automotive underhood | -55 | +125 | 60 &100 &140 | 1 1 1 2 | 1000 1000 300 40 | ~5 |

*$\Delta T$ represents the maximum temperature swing, but does not include power dissipation effects; for power dissipation calculate $\Delta T_j$; power dissipation can make pure temperature cycling accelerated testing significantly inaccurate.

250

**FIGURE 10.1.**    Handheld scientific calculator. [1]

## 10.1.1  Case Design

The continual packaging challenge created by products such as handheld calculators is providing more functionality in smaller packages. Thus, it is often necessary to integrate more than one function into a component. This helps to minimize the product's volume and helps to decrease the number of parts to be assembled. For example, it is not uncommon for the bottom case to provide the cosmetic and protective shell, and also to support the keyboard assembly.

For cost-effectiveness, heatstaking is a proven manufacturing process that can be used to provide uniform keyboard support. Using this process in combination with the case assembly eliminates the need for screws. Heatstaking can easily be automated because it is a machine-controllable process that requires fewer parts to assemble, and thus produces a sturdier product.

The cost-effective use of a molded thermoplastic for the case was also chosen in order to provide a consumer-pleasing clean appearance by keeping the package simple and free of overlays. A polycarbonate resin was used to help ensure that the product would survive some environmental shock, i.e., a one-meter (39 inch) drop.

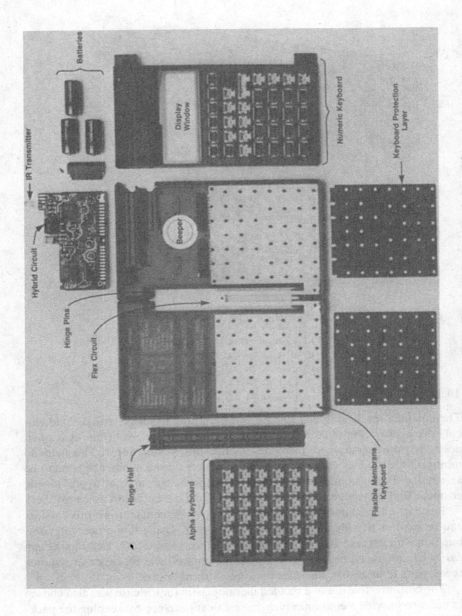

**FIGURE 10.2.** Exploded view of a handheld scientific calculator. [1]

## 10.1.2   Dense Packaging

A hybrid, double-sided, mixed-technology, printed circuit assembly, Figure 10.2, is often used in products to optimize the use of sophisticated, high-pin count surface-mount integrated circuits with conventional through-hole mount discrete components. The size of the substrate used is generally minimized in order to help to minimize the cost of the assembly and to minimize the space required to package it within the calculator. In addition, such assemblies are required to have the capability to interface with the key lines, a beeper, battery spring, and lines to the flat-panel display.

## 10.1.3   Display Interconnection

It is not uncommon to have the flat-panel display interconnect directly with the printed wiring board assembly by the use of low-profile, solderless elastomeric stacking connectors. To establish and maintain proper registration between the display and circuit assembly the position of the display is usually determined optically. The display can then be secured in a stainless-steel display clip using double-sided, pressure-sensitive tape. The display clip is then positioned into the board assembly using holes that are precisely located with respect to the display pads. Such assemblies, can be subsequently crimped together after testing.

## 10.1.4   Hinge Link

In the example shown in Figure 10.1, a compound hinge is used to connect the two halves of the case. This allows the product to be used in different positions throughout 360 degrees of rotation.

Several methods are possible for fastening the two hinge halves together. These include gluing, ultrasonic welding, heatstaking, and the use of threaded fasteners. Even though it requires the use of more complex tooling, a snap-fit design is often preferred because it offers the most repeatable simplified assembly process.

For optimum effectiveness, when designed properly the hinge on which the link rotates can perform several other functions. When they are conically tapered they can provide axial self centering on the hinge piece in each case half. With the tip of the hinge pin designed to preload against the inside of the hinge, it is also possible to create a functional drag that provides a high-quality feel when the product is rotated.

## 10.1.5   Keyboard and Flexible Circuit Interconnection

A technology that can be used for the integrated keyboard and flexible circuit interconnection is a conductive silver ink screened onto a polyester-film substrate. This approach allows a single substrate and screening to be used for both keyboards and the flexible interconnect, thus improving the reliability of the system.

Controlling the motion of the flexible circuit can help to minimize the locations of stress concentration. This aids in controlling the induced torsional stress.

For optimum cost-effectiveness, the two keyboards can be an integral part of the flexible circuit, using the same screened conductive ink for the keypads and circuit matrix. This is the technology typically used for membrane-switch keyboards. Tactile feel for the keys can be provided by two separate dome sheets of formed polyester. A spacer layer can support the domes while also providing a vent at pressure extremes and whenever the dome is actuated. Such layers are all attached to create a single part for ease of product assembly. It is not unusual for such consumer products to be tested for one-half a million key cycles in order to assure no electrical failures and minimal degradation. Life testing is also done under environmental conditions of both high-temperature, Table 10.1, and humidity. Thus, several iterations of key design and testing are often required to achieve the desired life and tactile feel.

### 10.1.6    Electrostatic Discharge Protection

The performance of electrostatic discharge (ESD) testing on most types of electronic equipment is a consistent challenge with respect to releasing the product design for production on schedule. This is because such testing typically cannot be performed until late in the project because a completed product is required. Thus, modifications (fixes) that might be required as a result of ESD testing often do not have time to be integrated properly into the product. With this in mind, it is often necessary to make special considerations for ESD protection early in the design phase of the product. One approach is to arrange for the fabrication of a prototype model that is built using existing components on a prototype printed circuit assembly. The results of ESD testing of this model will, if necessary, allow for the making of timely modifications to the design of the product prior to the production release.

## 10.2    DATA PROCESSING EQUIPMENT (COMPUTERS) [2]

Many factors must be taken into account when designing data processing equipment, i.e., computers. One of the primary challenges faced by the computer packaging engineer is improving reliability while reducing equipment cost and size.

## 10.2.1   Reliability

The reliability of a computer is measured by the failure rate of its individual subassemblies. This failure rate is normally expressed as percent failures per one-thousand power-on hours. To complicate matters the functionality of computers is increasing rapidly. This has been brought about by the ever increasing integrated circuit processor performance, circuit speeds, and integration levels. Thus, integrated circuit packaging characteristics have changed significantly over the last several decades.

For a given level of integration, higher device package terminal counts and performance demands have had to be satisfied, along with power and cooling requirements, prior to optimization for lowest cost. This has resulted in significantly higher interconnection wiring densities and power distribution levels. Associated with these improvements has been the ability to reliably manufacture and assembly multilayer boards with smaller conductor widths, closer spaces, smaller holes, and more layers. This, in turn, has caused packaging costs to experience startling increases in unit values, although when overall packaging density is compared, such developments can be seen to be justifiable and cost-effective. Collectively, all of these factors have placed major demands, from a reliability standpoint, on the packaging methodologies being implemented.

## 10.2.2   Subassembly Packaging

The basic approach being taken dictates the reliability risks to which the packaging engineer is exposed. The types of subassemblies and their size, the number of component risk sites, and the density all contribute to the overall equipment failure rate.

Reliability may be enhanced by packaging approaches that reduce failure rates, such as by substituting mechanically separable connector terminations with semipermanent soldered connections. Thus, having fewer, but larger and more sophisticated, printed circuit board subassemblies in a piece of equipment is definitely a step in the right direction. In fact, from a reliability viewpoint, a one-board system is the optimum. However, large subassemblies by themselves do not guarantee achieving a better than normal reliability.

Having many, if not all, of a computer's data processing functions on a single subassembly implies having significant increases in on-board signal communications and power distribution within one component mounting and interconnecting substrate. The fabrication of such substrates with features that are smaller and closer together may introduce inherent reliability problems due to the larger number of risk sites. Managing the increased thermal densities that are also likely to occur creates other packaging challenges.

### 10.2.3   Environmental Stability

Due to the wide range of uncontrolled environments that computers can be exposed to, the packaging engineer must give serious consideration to environmental stability when designing these products. For example, large temperature differentials associated with environments that are not air-conditioned places additional constraints upon the design and influence the failure rates. Solder joints are particularly susceptible to damage (fatigue) caused by large and frequent temperature changes.

High operating temperatures, combined with high relative humidity raise the possibility of insulation-resistance failures occurring at subassembly risk sites, especially in the presence of bias voltages. Also, moisture may interact with minute amounts of contaminants, either present in the air or left on components during processing, to produce chemical compounds that eventually will become conductive paths; in some cases such chemical compounds may be corrosive. If these compounds settle on electrical contacts they may create high-resistance connections or open circuits. In such cases, it is often necessary to conformally coat the printed circuit board assemblies.

The computer vibration environment may be separated into three distinct categories:

- High-level—low-frequency range vibration present during shipping
- Transient—vibrations that are present for short periods which are caused by equipment or vehicles operating near the machine
- Continuous-level—vibrations that are present throughout the life of the product which are normally caused by the computer itself, or by the computer's components such as cooling fans or peripheral equipment.

Such vibration (and shock) loadings must be dealt with in any computer packaging design. The most severe shock environments that a computer will experience are usually caused by human intervention. That is to say, handling during shipment or relocation, customer engineering operations performed on the product, and operator-induced shocks (such as bumping the enclosure), all must be taken into consideration by the packaging engineer.

### 10.2.4   Cost Considerations

After marketing specifications and circuit technology capabilities are defined, the packaging of computers are very dependent on satisfying cost-to-performance trade-offs. The definition of the lowest cost packaging approach becomes an integral part of this total system optimization effort.

In order to better understand the impact of packaging on total system cost, a breakdown of the major contributing factors must be developed. For com-

puters, the major groupings of system hardware, plus assembly and test cost
contributors are:

- Circuit logic field replaceable units (FRUs)
- Power and cooling
- Inter-FRU signal distribution, printed circuit boards and cabling
- System assembly and test
- Machine enclosure and accessories.

The circuit logic FRU portion is usually less than one-half of the total system
hardware plus assembly/test cost for a typical small size computer. With increasing levels of semiconductor chip integration, Figures 10.3, the packaging
portion decreases, as measured on a system basis, because the cost of integrated
circuits, memory devices, power, cooling, assembly, and testing remain relatively constant. This trend also continues for moderate increases in the number
of circuits per computer.

## 10.2.5  Functional Size

In general, the functional size of a computer is measured in terms of data processing rate, in millions of instructions per second (MIPs), or in number of

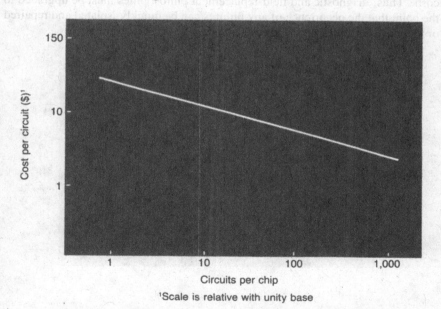

FIGURE 10.3.    Cost per circuit package as a function of level of integration. [2]

circuits. Strictly defining the data processing function as the main logic portion of the machine, i.e., the channel function and memory not included, the cost of circuit packaging on a per MIP basis has historically tended to decrease as the size of the data processor grew, Figure 10.4.

## 10.2.6  Other Considerations

Once the computer packaging approach to be taken has been defined to satisfy the basic performance, wiring, and functional needs, additional provisions must be made for manufacturability, testability, reliability, and maintainability. For example, system field maintainability philosophies will help to determine the basic choice of field-replaceable units.

As a packaging trend, the number of circuits per computer FRU has increased steadily from the mid-1960s, with from 5 to 10 circuits per FRU, to well over 10,000 circuits per FRU in the late 1980s. This increase is relatively much greater than the increase in the number of circuits per chip over the same time period. Hence, for most intermediate and high-end data processors, this reduces the number of subassemblies while significantly increasing component and FRU complexity. From a maintainability standpoint, this relates to having spare-part costs become a significant portion of the overall system life-cycle costs. Thus, diagnostic and field-replacement philosophies must be upgraded to the point that the occurrence of any failures can be quickly isolated and repaired or the use of throw-away modules can be cost-effectively justified.

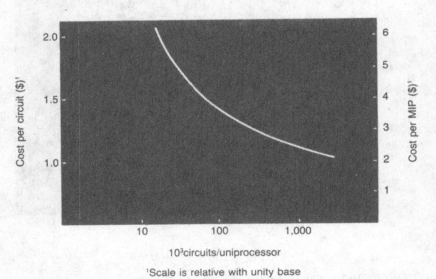

10³circuits/uniprocessor

¹Scale is relative with unity base

**FIGURE 10.4.**    Cost of circuit package as a function of size. [2]

## 10.3  AIRCRAFT (AVIONICS) EQUIPMENT [3, 4]

A major controlling factor in the packaging of electronic equipment for aircraft applications are the interface standards developed by the Airlines Electronic Engineering Committee of Aeronautical Radio Inc. (ARINC). These requirements are detailed in ARINC 404 and the ARINC 600 series of specifications. These documents provide for a system of modularized equipment dimensions with overall subassembly/system interface requirements, plus methods of attachment to mating mounting hardware and connectors, for use by avionic equipment packaging engineers. The details for the Air Transport Rack (ATR) system covered by the specifications have evolved over a number of years while acquiring increasing acceptance not just among the worldwide airline industry, but with the United States Department of Defense as well.

### 10.3.1  Case Designations

The ATR case sizes described in ARINC 404, Table 10.2, are designated with respect to a unit case width, i.e., 1 ATR or nominally 257 mm (10.1 inches). There is one standard height for all of the cases with a nominal dimension of 193 mm (7.6 inches) and two standard lengths, designated as short 320 mm (12.6 inches) and long 498 mm (19.6 inches). Examples of electronic equipment in ATR enclosures are the 1-ATR Short unit shown in Figure 10.5 and the 3/8-ATR Long unit (with access cover removed) shown in Figure 10.6.

The modular ATR electronic equipment racking system facilitates the mixing and matching of enclosure sizes. For example, a 1 1/2-ATR aircraft rack can, obviously, serve for the installation of one piece of 1 1/2-ATR wide equipment. However, it can also accommodate one 1-ATR and one 1/2-ATR unit, three 1/2-ATR boxes, etc. Of course, depending on the space available, all of the boxes on an individual rack shelf should be of the same depth, i.e., either all long or all short.

ARINC 600-7 also describes the dimensions and maximum weights for a series of LRU cases, Table 10.3. All of these cases are of the short variety and of the standard height.

For special applications there is a one-half height designation of a Dwarf ATR. Small electronic units with the facility for packaging equipment that is too small to fit into the standard ATR container are given the designation of Elfin Modules.

The standard height was originally designed to accommodate the vertical spacing of the horizontal rack shelving in airplanes. Therefore, manufacturers exceeding the standard incur the risk of finding that their equipment will not fit in the aircraft. However, when this height standard must be violated it is suggested that the tall enclosure be a maximum of 297 mm (10.6 inches).

**TABLE 10.2. Standard Air Transport Rack (ATR) Case Dimensions [3]**

| ATR Size | Approx Volume | | W | | $L_1$ | | $L_2$ (Max) | | H (Max) | |
|---|---|---|---|---|---|---|---|---|---|---|
| | In.³ | Liter | ±.03 in. | ±.76 mm | ±.04 in. | ±1.0 mm | in. | mm | in. | mm |
| Dwarf | 95 | 1.56 | 2.25 | 57.15 | 12.52 | 318.0 | 12.62 | 320.5 | 3.38 | 85.8 |
| 1/4 Short | 215 | 3.52 | 2.25 | 57.15 | 12.52 | 318.0 | 12.62 | 320.5 | 7.62 | 193.5 |
| 1/4 Long | 335 | 5.49 | 2.25 | 57.15 | 19.52 | 495.8 | 19.62 | 498.3 | 7.62 | 193.5 |
| 3/8 Short | 340 | 5.57 | 3.56 | 90.41 | 12.52 | 318.0 | 12.62 | 320.5 | 7.62 | 193.5 |
| 3/8 Long | 530 | 8.69 | 3.56 | 90.41 | 19.52 | 495.8 | 19.62 | 498.3 | 7.62 | 193.5 |
| 1/2 Short | 470 | 7.70 | 4.88 | 123.95 | 12.52 | 318.0 | 12.62 | 320.5 | 7.62 | 193.5 |
| 1/2 Long | 725 | 11.88 | 4.88 | 123.95 | 19.52 | 495.8 | 19.62 | 498.3 | 7.62 | 193.5 |
| 3/4 Short | 720 | 11.80 | 7.50 | 190.50 | 12.52 | 318.0 | 12.62 | 320.5 | 7.62 | 193.5 |
| 3/4 Long | 1120 | 18.36 | 7.50 | 190.50 | 19.52 | 495.8 | 19.62 | 498.3 | 7.62 | 193.5 |
| 1 Short | 975 | 15.98 | 10.12 | 257.05 | 12.52 | 318.0 | 12.62 | 320.5 | 7.62 | 193.5 |
| 1 Long | 1510 | 24.75 | 10.12 | 257.05 | 19.52 | 495.8 | 19.62 | 498.3 | 7.62 | 193.5 |
| 1 1/2 Long | 2295 | 37.62 | 15.38 | 390.65 | 19.52 | 495.8 | 19.62 | 498.3 | 7.62 | 193.5 |

NOTES:

1. The volume approximation given here does not include an optional front doghouse.
2. The $L_1$ dimension is the distance between the front surface of the designated front panel of the box and the surface indicated for MIL-C-81659 connectors.
3. All dimensions are overall including finishes. Projections such as screw heads, rivets, etc. are not permitted on the sides of the box.

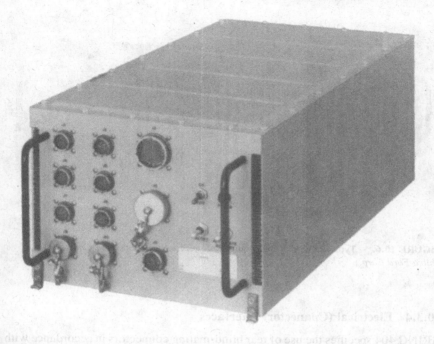

**FIGURE 10.5.**  Typical 1-ATR short equipment, with cold-plate cooling. (*Courtesy of EMM Corp.*)

## 10.3.2  Optional Projections

Projections up to a maximum of 63.5 mm (2.5 inches) beyond the front panel of the standard ATR enclosure are allowed in designated locations for the mounting of handles, connectors, displays, and controls as long as they do not interfere with the use of unit hold-down and extraction hardware. Projections on the rear of the case are permitted provided that they do not project 1.78 mm (0.07 inches) beyond the rear panel.

## 10.3.3  Mechanical Interfaces

The ATR units interface mechanically with the system hardware through the use of rear shock-pin blocks and front hold-down/extraction hooks. The details for the size and location of the features are clearly specified in the ARINC documents.

**FIGURE 10.6.**   Typical 3/8-ATR long equipment, access cover removed. (*Courtesy of Philco-Ford Corp.*)

### 10.3.4   Electrical (Connector) Interfaces

ARINC 404 specifies the use of rear blind-mating connectors in accordance with MIL-C-81659 or its commercial equivalents; ARINC 600 contains the details for connectors needed to satisfy high-density requirements. These connectors employ a rear release system for the loading, retention, and extraction of crimp-type contacts, but are also designed to accommodate nonremovable soldered and solderless wire wrap terminations. Configurations with a diverse array of shell and contact arrangements are allowed in order to satisfy a broad spectrum of avionics applications.

### 10.3.5   Air Cooling

Some aircraft racking systems provide for the forced-air cooling of the electronic equipment mounted on them. ARINC 404 describes the size and location of optional openings in the bottom of the enclosure for these installations. In other instances the cooling air can enter the equipment from openings in the front of the unit. In these applications air-cooled cold plates along the sides of the case, Figure 10.5, are used so that the cooling air (air that is sometimes contaminated) does not come into direct contact with electronic components within the housing. Another thermal management option is to provide heatsinks on the front of the ATR housing, Figure 10.7, that can facilitate the natural (free) convection cooling of the equipment.

TABLE 10.3.    Standard LRU Case Widths and Maximum
                Permissible Weights [4]

| LRU Case size | Width | | Maximum permissible weight Kg |
| --- | --- | --- | --- |
| | MM | Inch | |
| 1 MCU | 25.4 | 1.00 | 2.5 |
| 2 MCU | 57.2 | 2.25 | 5.0 |
| 3 MCU | 90.4 | 3.56 | 7.5 |
| 4 MCU | 124.0 | 4.88 | 10.0 |
| 5 MCU | 157.2 | 6.19 | 12.5 |
| 6 MCU | 190.5 | 7.50 | 15.0 |
| 7 MCU | 223.3 | 8.79 | 17.5 |
| 8 MCU | 256.3 | 10.09 | 20.0 |
| 9 MCU | 289.3 | 11.39 | 20.0 |
| 10 MCU | 322.3 | 12.69 | 20.0 |
| 11 MCU | 355.3 | 13.99 | 20.0 |
| 12 MCU | 388.4 | 15.29 | 20.0 |

FIGURE 10.7.    Typical 1/2-ATR short equipment, with heatsink cooling. (*Courtesy of Rolm Corp.*)

## 10.4  MILITARY SHIPBOARD ELECTRONICS [5, 6]

In the early 1960s, the United States Navy initiated a program to establish a family of standard electronic modules that, when properly combined, could make up most of the electronic circuitry required for military systems. Specific problems that required solution were:

- Inadequate reliability
- High life-cycle costs
- Long design-to-production lead times
- High maintenance skill levels
- High technical risks
- Poor logistic support
- Lack of a consistent approach to packaging technology.

Originally designated as the Standard Hardware Program (SHP), the project was later renamed the Standard Electronic Module (SEM) Program.

The purpose of SEM is to institute a module-level packaging standardization approach to the design, development, production, and logistic support of electronics equipment systems. The SEM Program provides the system developer with a family of commonly-used, low-cost, reliable, functional building blocks for the implementation of military hardware.

The goals of the program are to favorably influence system life-cycle costs, availability (reliability and maintainability), and supportability in order to minimize the impact of technology obsolescence. By specifying the functional, mechanical, and thermal interfaces for the SEM modules, current state-of-the-art and advanced packaging technologies and manufacturing techniques may be used cost-effectively. For example, SEM modules have been made both as printed circuit board and hybrid circuit assemblies and the most advanced versions are implemented with fine-pitch, double-sided, surface mount assembly technology.

### 10.4.1  Documentation

The SEM program is documented in a hierarchal structure of Department of Defense standards and specifications, including:

- MIL-HDBK-239—SEM Program Manager's Handbook,
- MIL-HDBK-1634—SEM Module Description Handbook,
- MIL-STD-1389—Design Requirements for Standard Electronic Modules,
- MIL-STD-1634—Design Requirements for Standard Electronic Modules
- MIL-M-28787—General Specification for Modules, Electronic, Standard Electronic Module Program.

**FIGURE 10.8.**     Typical size 1A SEM module assembly. (*Courtesy of Amperex Inc.*)

Well over 300 individual modules, such as the one shown in Figure 10.8, have been documented by the use of MIL-M-28787 slash sheets, i.e., detail specifications, that set forth their form, fit, and function requirements. SEM functional specifications not only encourage vendor innovation to provide a truly competitive procurement environment, but also provide a means for system technology updates at the plug-in module level.

## 10.4.2   Mechanical Configuration

In order to keep up with rapidly changing circuit technology, the packaging system chosen for the SEM Program is flexible and easily adaptable to many different component packages and interconnection media. Consequently, rather than selecting a single hardware configuration, the external mechanical interfaces have been specified while retaining an optional internal module configuration. In addition, the sizes of modules are allowed to vary in a controlled manner. This flexible packaging system has significantly contributed to the success and longevity of the SEM Program.

The basic size (Format A) SEM module, Figure 10.9, has a span of 66.5 mm (2.62 inches), a thickness of 7.37 mm (0.290 inches), and is 49.5 mm (1.95 inches) high, including keying pins. The module sizes increase in span, thickness, or both, Figure 10.10, while the basic height remains unchanged.

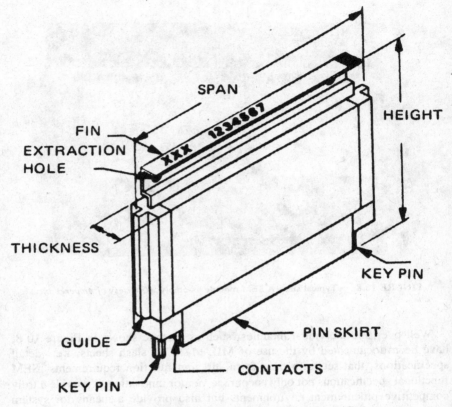

**FIGURE 10.9.**    Basic features of SEM modules. [5]

The basic SEM module span can increase twice in increments of 76.20 mm (3.000 inches). Thus, in addition to the basic span size (designation 1), there is a span size 2 of 142.7 mm (5.62 inches) and a span size 3 of 219.9 mm (8.62 inches).

The basic SEM module thickness can increase several times in increments of 7.62 mm (0.300 inch). As a result there are twenty-two allowable thicknesses beginning with the basic designation A through designation Z of 167.3 mm (6.590 inches). (The letters I, O, Q and X are not used.)

Any combination of the three spans and twenty-two thicknesses, such as the modules shown in Figure 10.10, can be used. This results in having the smallest (basic) module with the designation of 1A and the largest module with a designation of 3Z.

**FIGURE 10.10.**    Multiple-increment SEM module assemblies. [6]

### 10.4.2.1   Fin Structure

The fin serves as the insertion and withdrawal interface, identification marking surface, and as a means of module heat dissipation. Two holes are located in the module's fin in order to accommodate the use of an appropriate module extraction tool.

### 10.4.2.2   Guide Ribs

The guide ribs at each of the module spans serve to align the module during insertion into a backplane assembly. They also provide a conductive thermal interface.

### 10.4.2.3   Keying Pins

A keying pin is provided at each end of the basic module span to help ensure the proper mating of the module with the appropriate interface connector. These pins also help to align the module prior to engagement and serve as plug-in polarization devices.

### 10.4.2.4    Connector Pin Contact Skirt

The skirt provides a protective cover shield for the module's connector pin contacts. It also serves as a marking and identification surface.

### 10.4.3    SEM Connectors

Virtually all SEM modules utilize two-piece connectors with blade-type (male) contacts on the module that mate with tuning-fork type (female) contacts on the backplane. There are two rows of contacts in each connector on 2.54 mm (0.100 inch) centers within a row. The first variant is the basic SEM-A forty-position connector, i.e., two rows of 40 contacts each. Next comes the two-row SEM-B connector with 100 contacts for use with the double-span module. The SEM-C, D and E connectors have up to 400 contacts by having, when needed, a triple span or as many as four rows of contacts on centers as close as 1.27 mm (0.050 inches).

### 10.4.4    Improved-SEM Modules

A modification of the basic SEM family of modules is referred to as Improved SEM or I SEM. An ISEM module, Figure 10.11 differs from the basic SEM module in the following ways:

- The top surface has been increased in width in order to provide a better conductive thermal interface.
- The basic SEM fin has been eliminated, so that the circuitry is allowed to extend to the top of the module. Convection cooling is now accomplished by direct air impingement on the components rather than by passing cooling air through the fins.
- The guide ribs have been increased in both height and width in order to provide additional thermal-interface area.

**FIGURE 10.11.**    Leadless chip carrier ISEM assembly. (*Courtesy of RCA Corp.*)

- The center connector area has been filled in to provide a 100-contact capability instead of 80 (two times forty) for the basic double-span module.

These changes have resulted in significant module capability increases in component count, power handling capability, and thermal interface transfer. In all other respects the ISEM is completely compatible with the basic SEM packaging concepts, thus allowing them to be intermingled in new system designs.

### 10.4.5  SEM Thermal Management

In order to provide maximum flexibility for SEM users, the SEM Program thermally defines the module in such a way as to allow multiple system-module interfaces and cooling techniques. User data are furnished to assist the packaging engineer to optimally use the modules and adapt them to the equipment's particular thermal constraints. Within the module, as in the mechanical definition, thermal requirements are defined in a functional manner rather than by specifying structure, materials and processes.

The SEM Program defines a simple, clean-module thermal interface. That is, their packaging should be such that if the module's fin or guide ribs are held at or below 60°C., the module will meet its performance and reliability specifications. In addition, to facilitate system maintenance, the modules are required to perform if the fin or guide ribs are held at or below 80°C. (In both instances the lower temperature limit is 0°C.) This method of interface specification allows the module to be cooled by conduction or convection of heat from the fin, by conduction from the guide ribs, or by various combinations of these effects. Several systems use forced-air convection cooling, with or without the use of cold plates.

## 10.5  BIOMEDICAL ELECTRONICS [6, 7]

Electronic packaging is a vital technology in biomedical engineering. High reliability, small volume, and long service life (unique requirements for implantable devices) can be achieved only by the use the efficient high-density packaging of integrated circuits. Since a number of peripheral passive components are required to achieve the necessary circuit functions, the use of leadless, surface-mount, hybrid microcircuit technology is extremely important in fulfilling the rigorous size, weight, reliability, and quality requirements imposed on the finished product.

### 10.5.1  Electrical Considerations

The life-supporting function of most implantable devices, together with the requirement for low power consumption and low battery supply voltages impose additional demands on the implementation technology. Thus, the use of low-voltage CMOS integrated circuit devices has made a decisive contribution to the success of biomedical applications, although bipolar technology still has some biomedical applications.

The implantable devices in clinical use, Figure 10.12, perform complex physiological functions. Some of these functions are achievable only by using integrated-circuit microprocessors, memories, and corresponding peripheral circuits. Both linear and digital circuits are utilized, including operational am-

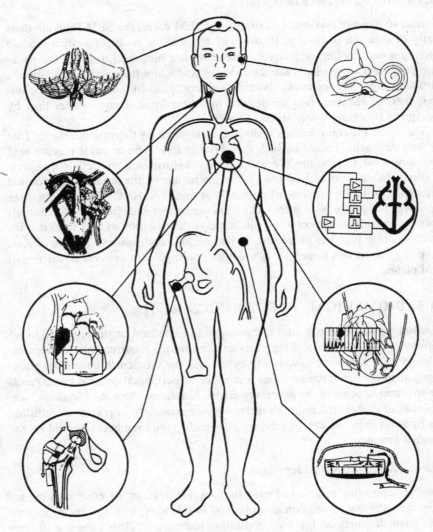

**FIGURE 10.12.**    Microelectronic implantable biomedical devices. [7]

plifiers, analog-to-digital (A-to-D) and digital-to-analog (D-to-A) converters, field-effect transistor (FET) switches, and transmission elements for the bidirectional transmission of operation parameters and physiological signals. The increased sophistication of implantable electronic devices has also caused significant increases in the amount of information being processed. As a result, the transition from analog to digital processing has become particularly important.

## 10.5.2 Microelectronic Characteristics

In most biomedical applications the following microelectronic characteristics are important:

- High reliability
- Minimization of power consumption, in order to achieve a service life compatible with the patient's life expectancy
- Extreme miniaturization in order to permit optimal placement of the device in available body cavities
- Short leadtime, between initial circuit design and the final packaged assembly
- Extreme versatility, in order to fulfill a variety of physiological requirements and internal operating conditions
- Hermetic encapsulation, of the circuitry and power source.

## 10.5.3 Cardiac Pacemakers

Microelectronics has had a decisive impact on the success of the cardiac pacemaker. Therefore, this device will serve as an example of the state-of-the-art packaging achieved for implantable devices.

Because of the life-or-death aspects of implantable cardiac pacemakers, high reliability is the pre-eminent goal of the electronic packaging of these devices. Product size/weight and competitive pricing come next in order of importance.

### 10.5.3.1 Types of Pacemakers

The cardiac pacemaker market is segmented into two general types of products. One is a single-chamber device that can sense and pace in the ventricle, the other is a dual-chamber device that senses and paces in both the atrium and the ventricle. The latter, is also capable of speeding up or slowing down its rate of pacing depending on the needs of the body.

Additional optional functions include:

- Programmable pacing through a range of different heat rates, adjusted through telemetry using a built-in wirewound antenna to sense external signals

- Internally sensing electrocardiograms (EKGs)
- Monitoring lead and battery impedance
- Permiting output parameters to be tuned in order to maximize longevity while still maintaining a good capture safety margin
- Keeping track of the number of times the device has paced in the ventricle or atrium;
- Storing information in memory in order to subsequently inform the physician how the patient and the pacemaker are interacting
- Help anticipate problems.

### 10.5.3.2  General Configuration

The basic elements of a typical modern pacemaker, Figure 10.13, are:

- CMOS-based VLSI integrated circuits
- Hermetically-sealed thick-film hybrid circuitry on a ceramic substrate
- A lithium battery
- A protective outer titanium case
- Leads that sense the electrical activity of the heart and transmit the pulses into the heart muscle
- A molded plastic cup to hold the circuit assembly and battery within the titanium case.

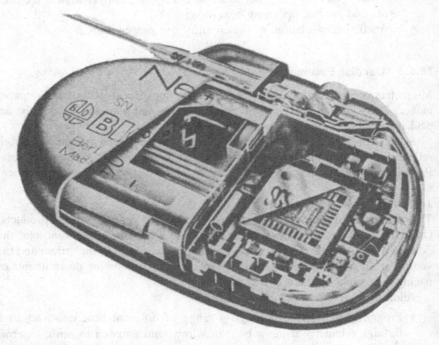

**FIGURE 10.13.**    Cutaway view of multiprogrammable single-chamber cardiac pacemaker. [7]

### 10.5.3.3 Packaging Implementation

There are many alternative approaches to both the packaging and assembly of cardiac pacemakers. For example, the integrated circuits can be either assembled as reflow-solderable, hermetically-sealed, surface-mount packaged devices or as a bare die that requires wire bonding or tape-automated bonding (TAB) and encapsulation. Also, some manufacturers use organic printed-circuit boards instead of ceramic substrates. Still others use co-fired multilayer ceramic substrates instead of traditional thick-film technology.

Cardiac pacemakers may or may not have a built-in antenna for external tuning purposes. When they do, in one approach the antenna is embedded in flexible circuit material in order to simplify and protect the connections.

The physician chooses the overall design size of the pacemaker for implantation using information based on the patient's age, size, and build. Thus, the pacemaker's case may be round, oval or rectangular. However, in general, single-chamber pacemakers are about 8 mm (0.3 inches) thick, while dual-chamber devices are about 12 mm (0.5 inches) thick. A round model is about 50 mm (2 inches) in diameter.

Electrical connections from the pacemaker circuitry to the outside of the titanium case go through glass-to-metal sealed feedthrough terminals. The lead wire that makes contact with the heart muscle is encapsulated in either polyurethane or silicone rubber, while the lead material itself is often stainless steel.

### References

1. Judith A. Layman, Mark A. Smith, Hewlett-Packard Co., "Mechanical Design of the HP-18C and HP-28C Handheld Calculators," *Hewlett-Packard Journal*, August 1987, pp. 17–20. (Copyright 1987 Hewlett-Packard Company. Reproduced with permission.)
2. John T. Kolias, Dean W. Masland, Robert L. Weiss, IBM Corp., "Cost and Reliability Drive Computer System Packaging Design," *Electronic Packaging & Production*, October 1986, pp. 80–83.
3. "Air Transport Equipment Cases and Racking," Aeronautical Radio Inc., ARINC 404A, March 15, 1974.
4. "Air Transport Avionics Equipment Interfaces," Aeronautical Radio Inc., ARINC 600-7, January 12, 1987.
5. David M. Reece, Ronald H. Huss, United States Navy, "The Standard Electronic Modules Program," *Electronic Packaging & Production*, July 1980, pp. 275–286.
6. M. Schaldach, Friedrich-Alexander University, "Hybrid Microelectronics in Implant Technology: a Progress Report," *Hybrid Circuit Technology*, April 1987, pp. 19–26.
7. Robert Keeler, Associate Editor, "Packaging to Pace the Heart," *Electronic Packaging & Production*, July 1986, pp. 28–30.

# Index